# ETHEREUM

The Complete Investing Guide in the Cryptocurrency Ethereum

(The Complete Step by Step Guide to Blockchain Technology)

**Vickie Marshall**

Published by Tomas Edwards

**Vickie Marshall**



*Ethereum: The Complete Investing Guide in the Cryptocurrency Ethereum (The Complete Step by Step Guide to Blockchain Technology)*

ISBN 978-1-990373-65-7

Legal & Disclaimer

The information contained in this book is not designed to replace or take the place of any form of medicine or professional medical advice. The information in this book has been provided for educational and entertainment purposes only.

The information contained in this book has been compiled from sources deemed reliable, and it is accurate to the best of the Author's knowledge; however, the Author cannot guarantee its accuracy and validity and cannot be held liable for any errors or omissions. Changes are periodically made to this book. You must consult your doctor or get professional medical advice before using any of the suggested remedies, techniques, or information in this book.

Upon using the information contained in this book, you agree to hold harmless the Author from and against any damages, costs, and expenses, including any legal fees potentially resulting from the application of any of the information provided by this guide. This disclaimer applies to any damages or injury caused by the use and application, whether directly or indirectly, of any advice or information presented, whether for breach of contract, tort, negligence, personal injury, criminal intent, or under any other cause of action.

You agree to accept all risks of using the information presented inside this book. You need to consult a professional medical practitioner in order to ensure you are both able and healthy enough to participate in this program.

# Table of Contents

# Introduction

This book contains a detailed but brief and introductory guide to Ethereum, one of the most popular and most profitable alt-coins at the time of writing. It is second only to Bitcoin (and the recently-created Bitcoin Cash or 'Bitcash'). However, there are some significant differences between Bitcoin and Ethereum, and it is the task of this book to explore some of the key features of this useful and interesting crypto-currency.

Cryptocurrencies are based on the concept of encryption – the idea that you can take information, encrypt it using complex programming, and then send it anonymously through the internet. They, furthermore, rely on the concept of the blockchain, which was pioneered by Satoshi Nakamoto, the pseudonym of the creator and founder of Bitcoin. Traditional currencies, called fiat currencies, still dominate the world of business and commerce, but cryptocurrencies are slowly catching up. Just as entertainment was revolutionized by the radio and television, and now the internet, so too is the world of currencies, trade and investment being shaken up by blockchain technology. If you are reading this book, or just glancing through this introduction, then you are clearly interested in cryptocurrencies and curious about Ethereum. Ethereum offers a tremendous opportunity for investment, as it is both cheaper than Bitcoin and innovative in its design. While this book does not constitute investment advice, and everyone should be cautious when investing in cryptocurrencies (never invest more than you are willing to lose), this book will provide a great introduction to Ethereum for the newbie investor or cryptocurrency enthusiast. In this book, you will learn:

The history and background of Ethereum

How Ethereum works

What smart contracts are

What the Ethereum Virtual Machine is

More information about Ether and other cryptocurrencies

How to invest in Ethereum

And some bonus information about Ethereum

Welcome to the amazing world of cryptocurrencies and blockchain!

And thanks again for downloading this book, I hope you enjoy it!

# Chapter 1: Ethereum Mining

Ether mining is a way of accruing ETH tokens through validating transactions on the network. Specifically, it is more participation in transaction validation and this is done to make sure that all activity that happens in a blockchain is confirmed. This can be done on all platforms, ensuring that is available to all, including those on home computers. That said, it is generally easier if you have a UNIX machine when you get started, rather than Windows.

The biggest challenge in Ethereum mining is the generation of more money through ETH collection than is spent on mining it. As a beginner to ETH mining, you are better off taking part in a mining pool, rather than trying to go it alone. You won't get so much to start with but you also stand a better chance of not being pushed out by professional miners. There are a lot of mining clients that can help you to get into ETH mining, ranging from small ones that are provided by CLI tools and core software to the more professional GUI apps that give you a clear view of all the activity and allow you to fine-tune your mining.

**Hardware**

Ethereum has now implemented The Casper Proof of Stake Algorithm and this means that any ASIC hardware that is used effectively for Bitcoin mining, and any other cryptocurrencies that are based on Proof of Work cannot be used for Ethereum any longer. Because of this, ETH mining is, right now, limited mostly to GPUs and this will exclude those Bitcoin ASICs that have limited the entry-level mining payoffs for hobby miners. Interestingly, it also favors the home consumer over the bigger investors. Either way, a similar scenario has raised its head where capitalism is continuing to manifest itself through the high-stake miners who have the means to purchase more equipment than beginners and hobbyists can.

To build your rig for ETH mining, you will need:

A motherboard, to allow everything to communicate properly

A graphics card, to process the algorithm for proof-of-stake

Some form of storage, be it SSD or HDD, for storing the Blockchain and any verified transactions

A decent amount of RAM, to provide memory for the mining program – be generous here, don't buy 2 GB when you can get 16 – things are only going to get bigger

A power supply

Ethernet, for receiving the validated transactions that are being stored in the Blockchain, plus all new ones that are validated while you are mining

Be aware that your graphics card will have the most vital role in determining whether your rig is lucrative or not.

## Mining Pools

Joining a mining pool will increase your chances of getting ETH tokens significantly. The reason for this is that the allocation probability, just like it is with other cryptocurrencies, is in proportion to productive across the whole network. By joining a pool, you raise your chances and any revenue received by your pool will be split between the pool participants where there are varying distribution agreements between pools. Type of payout includes PPS, or Pay Per Share, and PROP, or Proportional Payouts. As well as the pool, you will also need an Ethereum wallet, which is where your payments will go, and, of course, the mining software.

## Cloud Mining

Cloud mining is ideal for newbies to Ethereum mining because it implies that a service provider has packaged up and manages mining facilities that already exist. There are lots of different types of cloud mining, the most typical method being the purchase of tokens that are site specific, representing the right to a specified amount of hash power. This power is the mining power used by the service to mine the ETH tokens on your behalf. Other cloud mining types include:

Hosted – a service that leases machines out to clients

Virtual hosted – virtual server providers that lease the memory and processing power for general purpose and which can be used for ETH mining

Leased Hash Power - a provider that leases the hash power that allows the client to collect the profit

Mining Profitability

The profitability of ETH mining depends on a couple of factors – electricity, and hardware. The returns can be wildly varied and, as the currency decreases, the revenue can also decrease relative to your hardware and power costs. Cryptocurrency exchange rates are highly volatile by nature but, if you use the right hardware and the right energy source, you can put yourself on the path to earning a decent income.

## Mining Algorithm

Ethereum does not use the Proof of Work algorithm that Bitcoin uses, rather it uses Casper Proof of Stake and this wastes a lot less in terms of computational power, achieving the same result as Proof of Work at the end of the day. Casper prioritizes speed or availability over consistency and this means that it is more reliable and validations are faster. Where PoW uses hardware to provide a computation value, PoS takes a different approach, that of a miner-less direction, getting rid of the incredible energy requirements on the network. Instead of being reared for mining, Ethereum miners are given rewards that are in proportion to the transaction that they validate.

# Chapter 2: The Technology Behind Ethereum

Ethereum is based on blockchain technology. Blockchain technology is a kind of decentralized, public, and distributed ledger that records all transactions. It also has a high security. It is composed or records called **blocks**. Every new block that is added to the chain is connected to a previous block. This makes all the blocks interconnected with one another. The blockchain technology, or simply **blockchain**, is spread over a vast network of computers. Once a block is added to the chain, there is no way for it to be changed, modified, or altered, without changing or affecting other blocks. For any amendment to take place, there has to be consent from at least 51% of the users in the network. This is known as the **51% concept**. According to this concept, for any attack against the blockchain to be successful, it must possess at least 51% of the hash rate of the entire blockchain network. Since the network is spread over many computers/users, this is virtually impossible. Hence, blockchain has a high level of security. Take note that the 51% concept refers to the success of an attack and not to the possibility of being attacked.

It is worth noting that the Ethereum blockchain is the platform. It is powered by its token known as **ether**. The Ethereum blockchain is not just any other kind of blockchain. It also promotes the use of smart contracts and distributed applications. It is the presence of distributed applications that allows it to interact with the blockchain using smart contracts. The smart contracts run on every node of the Ethereum blockchain, which makes the application to be distributed.

There is also what is called as the **Ethereum Virtual Machine** or simply referred to as EVM. The EVM is the environment that runs and processes smart contracts. All the nodes in the Ethereum network that has a smart contract are run by EVM. It allows transactions to be made and actions to be automatically executed on the Ethereum blockchain.

In an Ethereum transaction, there is also what is called as **gas**. This term simply refers to the internal pricing for processing a transaction or contract on the Ethereum blockchain. Keep in mind that the Ethereum blockchain is powered by **ether**. The gas simply refers to the amount of ether that is needed to process and complete a transaction. It is often simply referred to as a transaction fee.

## Chapter 3: Trading And Mining

Smart Contracts - Smart Contracts

Smart deals represent the most interesting concept and at the same time perhaps the most important function enabled by blockchain systems. In 1994, legendary cryptographer Nick Szabo suggested that distributed books could be used for smart contracts. Nick Szabo and his texts were supposedly used as a role model for Satoshi.

Smart deals are automatic, self-executing, blockchain or digital contracts. Classic contracts are converted into program code, distributed and stored within the entire network. All devices that accept all conditions defined in the program code are kept and executed by them. Nick Szabo has just offered to date the best comparison of a smart deal. He compared this technology with a candy machine. When we put in a certain amount of money and put in candy to the appliance, it automatically throws out a candy swab.

In everyday life, we have two contracting parties that agree on some conditions. If both parties comply with the requirements, the contract is in force. Otherwise, if it is breached, the contract is terminated. Contracts are legally binding, and the foundations are the organization and functioning of human society. The most important progress blockchain brings the way in which it eliminates intermediaries even in contracts, with greater security. They, like classic contracts, define rules, penalties, and automatically enforce obligations.

When a smart deal is set up on a blockchain, each participant adheres to the terms defined by him by his own participation in the network. Conditions are no longer legally binding but are binding on the algorithm. To put it simply, a signed agreement from several parties will be unquestionable and automatically executed. It also covers the highest value of smart contracts - agreements, business rules and defined relationships are no longer stored in a separate database or a safe, or at some location already become life, an automated algorithm.

These characteristics of smart contracts allow their application and automation in many social and business situations and relationships. They contribute:

• Confidence - data on the blockchain is forever online

• Transparency - all parties to the agreement have committed themselves to clear initial conditions and can't claim otherwise

• Security - almost impossible to change because they are cryptographically protected

• Speed - automatic execution of transactions and contract conditions can't be compared with manual processing of documents

• Lower prices - if they allow the elimination of a plethora of intermediaries, create a space for great savings

- Precision - there is no longer a manual entry of data into the form, and therefore a space for errors is reduced

- Autonomy - consent depends only on the contracting party, and there is no longer a need for lawyers, brokers or similar mediators to confirm.

Digital Resources - Digital Assets

We had to mention previous concepts and Ethereum in order to discuss the topic of digital assets briefly. Many blockchain projects aim to bridge the gap between digital solutions and the real world. Products, goods, assets can be transferred to the blockchain network using digital certificates and smart contracts with all legally binding elements.

The digitalization of funds opens the door to various applications: the real estate industry, large markets for the sale, production chain management, and supply chains ... At present, it's probably impossible to imagine how much our reality will change this technology in the context of speed, communication, data transmission through automation and the most basic everyday things We deal.

The Ethereum is the third-largest crypto-market capitalization that exceeds $ 80 billion. It cost only $8 at the beginning of last year, and it has risen almost 100 times at the end of 2017.Growth in 2017: from $ 8 to $ 750, an increase of 9500 percent. However, they're managed to overtake another record, so this digital currency is close to 1,000 dollars. More precisely, it worth $ 1009.

Ether zero is the solution to stabilize the market. It will provide detailed technical plan change in the future. The people think it will change the crypto market that the world ever had. Ethereum can go to the moon. If you have there in the wallet, you will get free coin named other zero. It is a new launch coin. You have Ethereum then transfer to the portfolio. Its hard work supports and receives the free extra medal.

I will explain what will happen when you have coins in hard work. In the block, you had transactions you received 100 Ether. In the second you received 200 Ether and 100 Ether in the third block. After the third block, you have 400 coins. What it happens later? I can see that we did copy/paste. Now you have 400 coins in the first blockchain. It is an original blockchain. But you also have 400 in the new blockchain. All transactions are not visible in the other blockchain.

After 18th January Ethereum can hit $1800 in February. The second target would be $2000, and the third goal would be $5000. We can see clearly in the image below supported by a rising trend line. The price will continue to move to these three targets.

Yesterday's price movements

Today's price movements

Short term indicators showed buy direction 80%

Medium-term indicators showed buy direction 100%

Long-term indicators showed buy direction 100%.

Ethereum the third largest capitalization market

Trading Ethereum using MT4 trading platform and charts

Web trader platform selling position

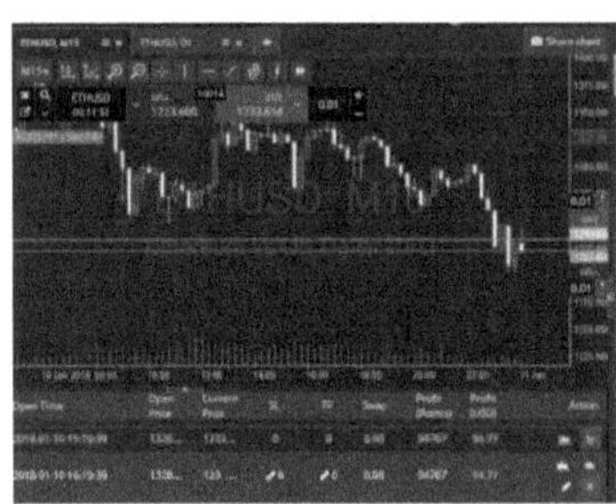

Trading history

Buying position

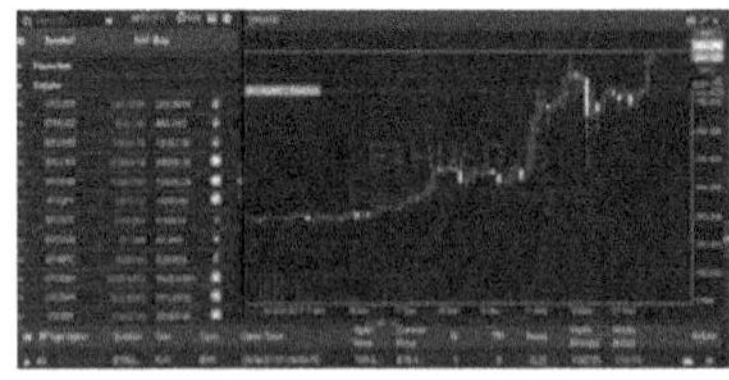

The whole world was affected by the impacts of Ethereum mining. For months, the AMD Radeon RX 570 and RX 580 graphics cards (as well as the older generation RX 470 and RX 480) have been sold out and can only be purchased at high prices. As the AMD card disappeared, more and more miners were moved to the GTX 1070, and GTX 1060 and GTX 1050 Ti series Nvidia GeForce cards, but now is GTX 1080 and GTX 1080 Ti. GeForce cards typically have a slightly lower digestion rate, but their power consumption is lower, and the total ratio is invested / even reached on the Nvidia card side.

As a rule, the largest earnings from mining are realized before the news about the sudden growth of the currency is widened. Miners who had the luck or intuition to start digging the ETH in 2016. they have grown over 20 times. As the currency is popularized, so more and more digger machines and the total weight of digging increases, which reduces the amount of excavated currency. Ethereum blockchain also has built-in difficulty-bouncing mechanisms when weight gains by 20% at one point. While calculating that the value of excavated currencies is greater than the cost of digging (electricity + repayment of machinery) mining is profitable at least marginal. It should not be forgotten that all calculators show current weight gain calculations, not weight projection and that earnings will surely be far less than in the event of a sharp rise in the value of the currency. But if the value of the currencies increases, this will re-popularize mining, and weight will grow rapidly.

Decentralized system

The specificity of the Ethereum is that it has the support from the creation, which, unlike the one that characterized Bitcoin while it was originally, is not only unique to enthusiasts, but also to technology and financial companies, where Microsoft is particularly interested. This technology of the future is based on the blockchain platform, which uses Bitcoin transactions, with significantly faster growth in the volume of representation, as well as with the application of innovative solutions that overcome the problems encountered so far in the cryptographic world.

The idea of the author is based on the intention of creating a single decentralized global computer that could completely transform the internet and its capabilities, adding to it, in financial operations, technical possibilities for money-laundering operations.

The digital currency called ether since 2015 represents a part of this network project and is intended to realize smart contracts and develop mutual trust among its users, which excludes the need for engaging mediators in transactions. The role that the digital currency ether has at the same time implies that it serves as a gateway for integrating applications into the network, but also as an additional motive for its owners to use the "mining" technique to maintain the vitality of the Ethereum network on which it is based.

These applications are specifically designed for the blockchain platform, which encompasses an enormously powerful common global infrastructure that can enable value changes and make an available digital property that is owned. This allows developers to work on creating new markets, storing debit and receivables information, and transferring funds based on pre-installed instances, in the way that futures contracts work in the capital market, as well as for many other functionalities that are still in production and those that will only be created, all without the need for a mediator and without any risk.

It will cover in contractual relationships

Ethereum was created in an effort to improve the Bitcoin network, where there was only the ability to conduct transactions, while it was not adapted to other activities, such as those practiced in banking operations with money, including such savings, and the like. It is precisely this role of the Ethereum in the shortest time that attempts to form, on the basis of a decentralized network that exists in the crypts, options with the aim of ensuring that digital money is taken as a cover in contractual relations, without the need to express its value in world currencies. These activities even at a time when the crypts first showed the momentum act irrational, but now they are not only initiated by alternatives, but there is also a huge interest of subjects that are in the financial world established players, such as global banks and card companies.

## The interest of financial institutions

The expectation that financial institutions, such as JP Morgan, might be interested in investing in such technology in the post-launch Bitcoin phase would be inconceivable, but such entities are investing millions of dollars into the development of startup companies engaged in research into the Bitcoin phenomenon and by facilitating its simpler traffic.

It was precisely JP Morgan that enabled the development of Fintech to start Digital Asset Holdings in order to continue researching Bitcoin potentials. This global financial institution initially invested more than $ 7 million in the financial startup of Digita Asset Holdings, which allows this type of crypto-currency to diverse forms of financial transactions outside the sale and purchase.

JP Morgan practically disassembled its competitors by enclosing the crypto-field as one that has all the advantages, and one of them was achieved thanks to the launch of Blythe Masters, with which the job is worth tens of thousands of dollars.

A series of experiments in a closed environment

At this stage, blockchain is used to store unprocessed data from Bitcoin transactions in a way that they convert to unchangeable files. This model is based on a commonly formed database that can be adapted to differentiation while retaining all existing performance. Today, an enormous number of companies mainly operating in a closed environment experiment with the Ethereum, while the desire for investment occurs with the largest market giants, aware that major changes are now underway in the field.

Ether and Ethereum should not be perceived as a threat to Bitcoin and blockchain, but as an emerging system that corrects all the details that have been detected as limitations on, now much older, crypts. Both systems are created with the same goal - by building greater trust in the conduct of direct transactions and by creating alternatives to the existing system in which end users, regardless of the level of investment, are constantly in an incomparably larger loss that goes to transaction costs in relation to intermediaries who are just such the circumstances of the opportunity to earn a profit. Ethereum only approaches the providers and users of financial services.

Transactions of limited models

Transactions of limited models, such as those occurring in Bitcoins, are based on a specific method of contracting or forcing the modeling process, such as the Ethereum, which can perform general-purpose jobs on a blockchain. After the implementation and wider acceptance of the Ethereum, the assets that are the subject of the transaction will be able to become the subject of savings and investments in other forms of property, all with a significantly higher level of online security.

The underlying reason why this technology sees an advantage over the existing one is that there are no elements in the private chain of crap tools that fully define only what are the characteristics of the chain, without the possibility of being in some other place. It is precisely in this that lies all the potential of transactions in such networks, since, after the definition of value, opportunities for exchange, storage, withdrawal and allocation of funds will be opened without the involvement of physical money or any other measurement units in the process.

In other words, regardless of the current currency rate in which the values of digital money have so far been expressed, the system is designed to function with one another relying on one's own rather than on external values. Also, the great advantage is that when transferring assets there will not be a complex calculation, since the assets do not actually move from the chain, but only the base record changes.

As for digging crypts, you need a very specific machine, equipped with as many graphics cards as possible. At present, the most popular graphics card is the Radeon RX 470, due to the price-performance ratio, which can confirm the shortage in the market. On average, the computer required for good mining should have a motherboard with a large number of PCI Express slots (5 or 6 is ideal), a processor that is almost irrelevant (Celeron "ends" the job), and 8 GB system memory. Of course, without quality power, there is no speech because these graphics require a large amount of electricity. Because of the impossibility of putting all of them physically into one board, there are so-called "rays" that allow you to virtually raise graphics and get the ability to not connect them directly to the board. This allows for better cooling, but also more graphics on one board. You can buy cases for such systems as finished, and you can make them yourself according to your desires and needs.

Ethereum is an open-end software platform, with a smart contracting function, based on blockchain technology that allows developers to build and implement decentralized applications. It can also be used for codification, security, and trade of everything - voices, domain names, financial exchanges, the allocation of financial resources, company management, contracts, and agreements.

Like Bitcoin, Ethereum has also been distributed to the public blockchain network, although there are significant technical differences between them.

While the blockchain of Bitcoin is used to monitor ownership of the digital currency (Bitcoin), the blockchain of Ethereum focuses on starting the program code of any decentralized application.

With this blockchain, "miners" are working to earn an Ether, which is essentially the currency of the Ethereum platform. According to Etereum's founder, Ether is a type of crypto token that runs the network and is used by app developers to pay fees for transactions and services within the framework

Ethereum network. As with Bitcoin and Ethereum, it can be traded and invested.

After a hacker attack in which Ethereum banknotes worth $ 60 million was stolen, due to disagreement with Ethereum's operations, a minor change in the Ethereum code occurred, resulting in the Ethereum Classic (ETC) cryptocurrency.

Therefore, two identical competitive currencies of ETC and ETH (whose price is significantly higher and more present in the crypto community) have been created.

Characteristic of this cryptocurrency is a smart deal or intelligent contract technology. A "smart deal" is, in fact, a phrase used to describe a computer code that can facilitate the exchange of money, assets, contents, assets or any value.

When a blockchain starts, a smart deal becomes a type of standalone program that automatically executes when certain conditions are met. Because smart deals work on blockchain technology, they function exactly as they are programmed, without any possibility of censorship, delays, fraud, or fraud on the third party.

While blockchains are largely limited in code processing, Ethereum demonstrates its diversity by offering developers the ability to create the operations they want.

How to get Ethereum?

The easiest way to buy Ethereum (i.e., Ether) is through some of the Bitcoin exchange (exchange offices) that also offers Ethereum.

Coinbase is one such exchange that offers perhaps the cheapest and most appropriate way to buy Ethereum. All you have to do is open an account at Coinbase, add a payment method to your account (credit card or bank account) and option "buy/sell "(Buy/sell) select" buy ETH "- buy ETH. The fee that Coinbase charges vary from 1.49% to 3.99% depending on how you pay (credit cards have a higher commission).

Cex.io is also a Bitcoin exchange office where you can buy ETH through a credit card, and their fees are already included in the exchange rate (which makes it more expensive than others). You also need an account at Cex.io where you will add a payment method, then with the option "buy/sell "you choose ETH and the quantity you want to buy. For note, Cex.io is worldwide unlike Coinbase, which means that it is available to many countries around the world.

Coinhouse

The option available for EU users, which initially functioned only for French citizens, but quickly expanded to the rest of Europe. With this option, you can purchase ETH by credit, debit card or via Neosurf.

Places, where ETH is available for sale, are also Kraken, Poloniex, and Shapeshift.io.

If your goal is to profit entirely on the ETH course, you could also invest in Ethereum CFDs, which is essentially a contract for difference. The idea is simple and based on it instead of buying Ethereum; you can only trade according to the exchange rate of this currency. It is most suitable for experienced traders because money is at risk for these transactions.

Currently, Plus500 is the only company offering such contracts.

If you are interested in mining Ethereum, you can try it from your own computer (CPU mining), or to install a dedicated GPU for that task.

The choice is about how much of the "future of the internet" will be in your hands.

An example of the potential application of smart contracts is as follows: Person A has decided to order a particular product from person B. At the moment when both parties agree on the price of the goods and the date of delivery, person A will send a certain amount of money into the Ether on a specific Ethereum address. After that, this money is locked in the smartest contract itself and is awaiting further execution of the contract. If the goods arrive before a certain date, the smart contract will automatically transfer money to the person B who sent the goods. Otherwise, the person A will be returned the money back. Similar systems existed earlier with one difference; there was a broker who, as a rule, increased the transaction itself, often slows down the transaction itself and, most importantly, it is necessary that persons A and B trust the intermediary. With clever contracts, the agent is thrown out completely, the trade and the transactions are accelerating, the costs of the brokers no longer exist, and most importantly, it is no longer necessary to trust the intermediary or the person with whom we operate. The contract itself is defined by computer code, and no one will be able to modify it or affect its execution. The example I have stated is fairly simple and aims to bring the way to the functioning of smart contracts and to present the capabilities of this technology.

Potentially, the application of smart contracts can be significantly more complex. Areas, where the potentials of smart contracts are examined, are real estate, insurance, stock exchanges, voting, but also many other areas where there are intermediaries that introduce higher costs and longer execution of certain jobs. Benefits of decentralized applications and smart contracts in relation to existing applications are:

• Data stability. No one can alter any data because all data is stored in a blockbuster and is found on thousands of computers around the world.

• Application security. Since there is no centralized server on which the application is located, but it is distributed to thousands of computers, no one can hack the application and cause harm to the users of the application.

If you want to mine another coin, for example, Ethereum uses graphics cards, and cloud mining does not pay off, which means you need to configure the configuration yourself and everything else. If you already have a graphics card, the next step is to download mining software like Claymore Dual Miner, where you can at the same time mining Ethereum and another coin, such as Sia or Decree. After you open the miner, you will see the start.bat script in it, input your mining pool, I use Nanopool and your wallet address for the selected currencies. When you start, the process begins, and you will see who cash rate, so you can count on how much you will earn. For all calculations, use whattomine.com also here you can see which is the most profitable coin at the moment, for AMD cards, it's mostly Ethereum for Nvidia, it's Zcash or Zencash, but it's all that often changes. It all depends on the market and the price on the stock exchange, so it needs to be very careful with investments in crypts because they are very subject to sudden price changes, both higher and lower.

Here are some indicative figures for the ETH + Sia mining configuration with 4 AMD RX570 4GB cards, which is 120 MHz for ETH, you can make such a configuration from about $ 1500, and in a year, it can make ETH and Sia in a value of $ 4,000 which means In less than half a year your initial investment returns. To get the higher hash rate on your cards, you can overclock them with software like MSI Afterburner, for AMD cards it is required to install Blockchain driver, Bios mod, etc. When calculating profits, we have to consider both the consumption and the price of electricity that is 12 cents per KW / h in our country.

EthDcrMiner64.exe is used to start the digging program and therefore after timeout / t 60 is placed as the first parameter.

-the pool is a parameter that shows us on which pool we will dig. In this case, it is eu1.ethermine.org, and this pool is accessed via port 4444 (eu1.ethermine.org:4444). The address and port for access to the pool are different for each pool, and accurate information can always be found on the official website of the pool, usually in the help / how to connect sections and the like.

-Well is a parameter that shows us which wallet we will dig. A wallet can be your local (installed on your computer) or a wallet issued by a stock exchange that is otherwise not recommended because in this way there is a possibility of losing your fund if something bad happens to stock exchanges. Pools also give you the option to add the name of the rig with the wallet/wallet address in order to have more transparent statistics and separately see the hash rate of each of your rigs (<address_name>. <Name_riga>).

-PSW is a parameter that shows us which password we use to access the pool (default is left "x"). For some pools where worker/rigging is done, you can choose any password.

After you are well acquainted with the basic parameters for setting Claymore's Ethereum miner, you can find in the Readme file many advanced parameters that you can use with the basic parameters to gain better control over your hardware.

The Start.bat file of this other variant tells us the following:

Run EthDcrMiner64.exe

-pool dig a pool eu1.ethermine.org:4444,

-well everything is dug into the wallet (red text) 0xfb1eAb156bdc4F1F082864751F68E8B36463Bc31.RX580 and we'll see the rig on the pool statistics under the name RX580 (green text)

-PSW uses a letter x for a password

-wd 1 system for self-detection of software and hardware errors should be included

-r 1 if the system detects an error start reboot.bat to restart the system

-Fan min 65 The minimum fan speed on the cards is 65%

-Fan max 85 The maximum fan speed on the cards is 85%

-tt 56 maintain the card temperature at 56 degrees using the range of 65-85% of the fan speed

-stop 65 if, with the specified range of 65-85% of the fan speed, the temperature of the yarn is 65 degrees, the bath goes out

Note: to use the -r1 radio as described above, you need to create another batch file named reboot (reboot.bat) in the folder from the miner where it is located and start.bat, the following:

shutdown / r / t 5 / f

These are just some of the commands that help you achieve better performance and better control over your hardware.

In addition to the parameters listed in the second example, there are parameters for overclocking graphics cards, and you can find them in the Readme file located in the miner folder. Before you leave the rig unattended, I advise you to first carefully examine the entered parameters and make sure everything works properly.

For years, Ethereum has been one of the most popular and profitable coins for mining. To miner Ethereum and to make it profitable you need to have a mining rig with newer generation graphics cards (RX 470, RX 480, RX 570, RX 580) with a minimum of 4GB of RAM. By upgrading the digging software the adjustment process is considerably facilitated in relation to the first generation of the software, but many beginner miners still encounter problems when setting up. To release your mining rig to ETH miners, you need to download the latest version of one of the mining software, in this case, Claymore's miner. In the first post in the Claymore's Dual Ethereum topic at the bitcointalk.org forum, you can always find and download the latest version of Claymore's miner.

After downloading the latest version of Claymore's miner, unzip the archive (WinRAR document) to the location where it will be your most convenient. Before you start setting up Claymore's miner, you need to find the POOL you will use because SOLO mining ETH has long been unprofitable. On the website at www.etherchain.org, you can find a graph showing the most popular POOLs. To start the ETH mining process without errors, you need to set up the mining software correctly. The settings are done through the start.bat file. If your digging software starts automatically when starting/igniting the mining rig, it is preferable to use the timeout / t 60 option in the start.bat file that is placed at the start of the start.bat file to delay the start of the digging for 60 seconds for all important services and applications (MSI afterburner, Sapphire TriXX, GPUZ) needed to run mining hardware.

There are several ways to buy Ethereum, Ether or ETH (this is the brand name), and this guide will show you how you can keep someone in the easiest way. As with Bitcoin, there may be quite a few jump hoops, but we hope that we will show you the best way to adapt to yourself. You can buy it with a fiat currency, buy Bitcoin or miner it.

There are plenty of exchange offices where you can buy ethereal, or ether, by substituting for traditional local currencies such as dollars, euros, yen. If you cannot find the right currency for trading, or if you think that the distribution of the requested offers is too high or you do not have enough liquidity, you will have to buy it in some other way. The best way to do this is to buy Bitcoin first. (hyperlink to the post about the purchase of Bitcoin). Now that you have Bitcoin, you can get an ETH or an ether using one of the BTC-ETH-based money changers.

Another option to get an ether is through mining. Instead, you can try to buy a cloud mining contract with Hashflare or Genesis Mining. Editing can be a little problematic. We suggest you join a mining pool to reduce return volatility. If you want to buy a mining contract, that means you are transferring a small portion of your profits to someone else taking care of managing, maintaining and adjusting costs.

Ethereum was established as a decentralized and autonomous currency. Although it is not managed by the management team in the traditional sense, it is better structured than Bitcoin. In the past, the development team chose a one-way solution to critical issues. A recent example would be the way they reacted to the DAO attack. After the attack, 3.6 million ether was stolen. Ethereum had two options: 1. to leave things as they are or 2. remove the stolen currency from the blockchain. Developers made a new chain, hard fork, and removed the stolen Ethereum from the blockchain and allowed users to choose which chain to follow. Now there are two ether chains; the old one is also called ether classic. Ether classic is an original chain containing stolen ether, while the new one does not contain.

# Chapter 4: Difference Between Ethereum And Bitcoin

This chapter will help you differentiate Ethereum from Bitcoin because many people group them under the same category even though they are very different. In most cryptocurrency conversations, the two must be discussed due to their popularity, but people do not understand that they are very different.

Ethereum came into the limelight in 2014 when Vitalik Buterin announced it in a Bitcoin Conference in North America. One of the main reasons Ethereum has become popular is because people have compared it to Bitcoin. In the previous chapter, we talked about blockchains, and we said that its value depends on the type of data that it contains. We also noted that a blockchain has financial data that is comparable to dollars while Ethereum is a slow computer platform or technology.

The difference between the two is that Bitcoin is a currency (such as British Pounds) while Ethereum is a platform/technology that people use as a basis to build other different technologies.

Now that you have a better understanding, let's look at what Bitcoin is and how it differs from Ethereum.

Bitcoin

In 2009, Bitcoin was created by Satoshi Nakamoto, who wrote a white paper about it and introduced it to the world. The thing that makes Bitcoin a desirable form of payment is that it has lower transaction fees than most other online payments. Typically, when you make online payments via options like PayPal or wire transfers, there are some transaction charges that you would incur. These costs are relatively high as compared to what you pay when you use Bitcoin.

Another thing is that the government does not in any way control Bitcoin like it does other currencies. Therefore, this cryptocurrency has little to no formal regulations as other forms of money do. Standard currencies have many rules including fiscal and monetary policies from commercial banks, central banks, and financial institutions. However, Bitcoin does not work this way.

Since it is a virtual currency, you cannot have a physical Bitcoin. An easy way to understand is to think of it as a "digital dollar." In order trade with it, you have to create an account through a website so that you can buy and sell your Bitcoin. People can now change their Bitcoin and use them as tokens, which is a form of currency that allows people to invest in companies in the same way as you would when buying stocks in an IPO (Initial Public Offering). With the Bitcoin, this is called the ICO (Initial Coin Offering).

The reason people use Bitcoin is so that they can store their money as securely as possible. People see it as a very excellent investment option because of the steady appreciation in the price. Others even use it to raise funds through the ICO.

Ethereum

We already described Ethereum as not just a cryptocurrency but a platform that runs on contracts. Many people predict that it will eventually overtake Bitcoin, but what differentiates the two is the technology behind them.

Here are the main differences between these two cryptocurrencies:

**Economic model** – In terms of the economic model they use, the two of them are quite different. The Ethereum model releases the same amount of Ethers annually forever. Bitcoin, on the other hand, releases rewards for a long time, even up to 4 years.

**Definition** - Bitcoin is a cryptocurrency while Ethereum is a ledger technology system that users can build upon.

**Block time** – This refers to the time it takes for a block to form. For Bitcoin, it is 10 minutes, while for Ethereum it is 15 seconds.

**Transactions** – The unit for transactions on the Ethereum platform is Ethers, while

Bitcoin in itself is a currency. Bitcoins can be broken down into smaller units called Satoshi and Milibits, but Ether has different denominations such as Kwei, Mwei, and Babbage.

**Inception** - Ethereum came as a result of crowdfunding, but Bitcoin was something that was created by a group people and introduced to the world. These individuals own most of the Bitcoin supply.

**Terms of ownership of currency** - In Ethereum, there is likelihood that 50 percent of the Ethers will be owned by people in the future. With Bitcoin, most of the currency is held by the people who first mined the currency.

**Roles** - The two cryptocurrencies have different functions. Bitcoin is focused around the disruption of mainstream world currencies while Ethereum is about the disruption of equity.

Another point to note is that some Fortune 500 companies have come together to try and build upon Ethereum technology. They have liaised to support the Enterprise Ethereum Alliance, which is the Ethereum smart technology we talked about in the previous chapter.

These are some of the outstanding differences between these two major cryptocurrencies. To differentiate the two, you need to understand both of them as separate entities. This will aid you when you are researching further about different cryptocurrencies.

In the next chapter, you will learn about how to mine Ethereum and the process that you need to go through to create Ethers.

# Chapter 5: Ethereum Virtual Machine

A virtual machine was put into place for Ethereum so that security could be taken care of whenever code that cannot be trusted is executed since this piece of code is executed by almost every computer in the world. Whenever you observe the virtual machine, you will realize that the virtual machine will work on Ethereum's security against attacks; more specifically DOS attacks that are directed at cryptocurrency platforms. The virtual machine is also going to make it to where external programs are not able to interfere with any communication points that are running the program.

The chances are that you are not a programmer, so you are not going to know what a DOS attack is let alone how to prevent them. Therefore, you need to understand what they are and how the virtual machine will stop them from attacking the Ethereum system.

Ethereum virtual machine will run off a runtime environment as smart contracts are executed. And, with how popular smart contracts are becoming by Ethereum users, it is possible that these smart contracts will take over the financial industry. However, the smart contract technology will complete tasks that have to be achieved without having supervision which makes it a version of machine learning.

A paper was written by Dr. Wood that stated that the virtual machine was created in a sandbox environment which means that it will have the ability to change the future of cryptocurrency because there is one piece of code that will outperform every other platform.

The sandbox environment is not going to be the ideal environment because you are not going to be able to see the program's full potential due to the fact that the initial states continuously change. Sandbox environments are not going to be like the real world where the users are because the users will use the program in a way that is different than how a computer will use it. But, testing the program in a sandbox environment is one of the safest ways that developers will be able to check the constraints of a program without releasing it to the public. This means they can ensure that the coding for the program is right and is not going to crash on the users thus causing the users to get upset and possibly leave the program for good.

While you watch the day to day operations on a decentralized system, you will know that the virtual machine will be what is in charge of making sure that those tasks are completed in the order that they are supposed to be completed in.

The best thing about the virtual machine is that it is free! This means that every programmer will download it and use it.

# Chapter 6: How Cryptocurrencies Work: A Detailed Guide About Blockchain

In the last sections, we talked a lot about what cryptocurrencies are but very little about how they work. To make a sound decision, you need to know this because without knowledge of how cryptocurrencies work, you won't be sufficiently equipped to make the right choice.

How Cryptocurrencies Work

As we stated earlier, the underlying technology behind cryptocurrencies, the technology that gives them their applicable use, is the Blockchain. The Blockchain technology is a technology that supports multiple technologies (cryptocurrencies are just one) such as cryptographically secure voting, database maintenance, digital identity, and many others.

The Blockchain technology is synonymous with Bitcoins because in reality, while the technology has seen varied seen developments from different pioneers in this field, it is the ingenious invention of Satoshi Nakamoto, the pseudonym given to the inventor/s of Bitcoins.

Blockchain is the underlying technology that allows for the distribution of (not copying of) digital information. As the name suggests, a Blockchain is a chain of blocks linked together by cryptographic code. A block is a complex cryptographically secured, complex mathematical problem (called a "hash") that computers (miners) on the peer-to-peer system that supports the Blockchain seek to solve.

Once a miner (or group of miners) on the network solves the mathematical problem underlying a block, it completes the block. Once a block is complete, you cannot change the information housed inside that block without meeting specific conditions. Any attempt to change the information—information about transactions within that block—voids or breaks the block, which then affects all other blocks on the system.

When miners complete a block, they unlock a new one and for their effort at solving the block (not them per se, but their computing power) they receive some coins as remuneration.

To make the chain complete (to make the Blockchain), once created—which happens when miners solve the cryptographic code that secures the block—the new block takes data from the previous block thereby creating a chain. The most amazing thing about the Blockchain is that in an instance where someone changes some elements of one block, because any new block created takes some data from the previous block, the singular change topples over to all other blocks from the instance of change henceforth. This means older data on the Blockchain is more secure because we measure Blockchains using height, which is the number of blocks in a chain.

## About Mining

Through this section, I have mentioned various times how miners are responsible for solving the mathematical problems that unlock current blocks and create new ones. Well, the work of the miners is to enter, verify, and bundle transactions together so that they can solve the mathematical puzzle.

In reality and as we've hinted at several times in this guide, miners do not do this work by hand and if you want to become a miner, you don't need to be a mathematical whiz. Instead, they dedicate their computing power to the prospect. Because blocks are cryptographically secure, the computers used to "mine" new coins, which in essence means verifying and bundling transactions, have to be powerful ones. Such computers cost a lot of money and use up a lot of power especially in relation to the ever-increasing difficulty of solving the cryptographic puzzle that unlocks a block and new coins (in relation to cryptocurrencies).

As I mentioned, to incentivize miners and make sure the system has enough computing power—that is the power of distributed ledger and what makes the Blockchain such a powerful technology—the system rewards miners with transactional fees and new coins after they solve the puzzle that unlocks new coins (sometimes the system can reward miners with both).

If you do decide to become a miner, you will need to buy a mining rig—learn more about mining rigs here. A mining rig is an especially powerful computer created for the sole purpose of optimizing graphical power and mining capabilities. Part of its job (and your job as a miner) is to ensure the validity of transaction by ensuring that the person attempting to, for instance, sent cryptocurrencies, has enough balance. To do this, miners (the mining rig) examines the existing block and Blockchain to determine the balance within a wallet. This is especially useful because as you now know, the Blockchain (the distributed ledger) is publicly available and anyone can access it.

Because the ledger or Blockchain has every transaction conducted on the system and from wallet, you may be tempted to (and forgiven for) thinking that this compromises the security of the system since anyone can track how you spend cryptocurrencies and the balance in your wallet. However, that is not the case especially since most cryptocurrencies and applications on the Blockchain technology allow for complete anonymity while still maintaining the integrity and openness of the Blockchain. Further, you can have as many cryptocurrency wallets as you want.

When it comes to mining, if you look back at what we've discussed thus far, you will recall that to solve current blocks and solve new ones, miners have to solve a mathematical puzzle and add the hash to the chain. Because doing so amounts to adding information to the block and chain, when a miner successfully solves the puzzle that unlocks a new block, he or she has to announce the solution to the network.

The peers on the network then scrutinize the solution and if it's correct, accept the hash into the block and chain—we call this consensus, the purpose of which is to ensure rogue miners do not try to trick the system by facilitating invalid transactions and by so doing, create new blocks.

That's about it about how cryptocurrencies work and the nature of the Blockchain, the underlying technology driving the growth of cryptocurrencies and other Blockchain related technologies.

To compound this knowledge, watch this video:

With that out of the way, and because you now understand cryptocurrencies and their underlying technology, let's talk about whether they are right for you.

## Chapter 7: Ethereum And It's History

Ethereum was created with the aim of building software that will create numerous decentralized applications. Buterin had argued that Bitcoin needed a **scripting language** for application development. Failing to gain agreement, he proposed the development of a new platform with a more general scripting language.

Although he started the research alone, he was later joined by Dr. Gavin Wood who later became a co-founder of Ethereum. The original four members of the Ethereum team were Vitalik Buterin, Mihai Alisie, Anthony Di Lorio, and Charles Hoskinson. Formal development of the Ethereum software project began in early 2014 through a **Swiss** company, **Ethereum Switzerland GmbH** (EthSuisse). Eventually, a Swiss non-profit foundation, the Ethereum Foundation (Stiftung Ethereum) was set up as well. Development was funded by an online public **crowdsale** during July–August 2014, with the participants buying the Ethereum value token (ether) with another digital currency, **bitcoin**. While there was early praise for the technical innovations of Ethereum, questions were also raised about its security and scalability.

Several prototypes of the Ethereum platform were developed by the Ethereum Foundation, as part of their Proof-of-Concept series, prior to the official launch of the Frontier network. The last of these prototypes culminated in a public beta pre-release known as "Olympic”. The Olympic network provided users with a **bug bounty** of 25,000 units of ether for stress testing the limits of the Ethereum blockchain.

After Olympic, the Ethereum Foundation announced the beginning of the Frontier network to mark the tentative experimental release of the Ethereum platform in July of 2015. Since the initial launch, Ethereum has undergone several planned protocol upgrades called milestones, which are important changes affecting the underlying functionality and/or **incentive structures** of the platform.

The current milestone is named "Homestead" and is considered **stable**. It includes improvements to transaction processing, gas pricing, and security. There are at least two other protocol upgrades planned in the future, i.e. Metropolis and Serenity. Metropolis is intended to reduce the complexity of the EVM and provide more flexibility for smart contract developers. The move to Serenity is still uncertain but should include a fundamental change to Ethereum's consensus algorithm to enable a basic transition from hardware mining (**proof-of-work**) to virtual mining (**proof-of-stake**). Improvements to **scalability**, specifical **sharding**, are also said to be a key objective on the development roadmap.

Progress Made So Far

Even though still in its early days, Ethereum has already seen a number of projects emerge that are seeking to bring its core concepts to life. Far from just theory, Ethereum-based projects are inspiring developers, overcoming challenges in the wild, inspiring research papers, grabbing global headlines and operating without the backing of a conventional corporate structure. Let us now discuss some early and notable examples.

The DAO (Decentralized Autonomous Organizations) The most prominent ethereum project yet launched, The DAO was a DAO designed to collect ether investments and distribute those funds to projects voted on by an open community of donors and members. In its short lifespan, The DAO amassed upwards of $160m denominated in ether, and saw a number of proposals put forth for voting, though none were passed. The DAO quickly emerged as a magnet for academic criticism about how DAOs should be designed and their participants incentivized. At this time however, the project had effectively collapsed following an incident in which an attacker was able to exploit functionality in The DAO's code. Called a "recursive call exploit", the attacker effectively requested funds from The DAO repeatedly, and the contract approved these fund requests without first checking the balance. Presently, ethereum developers were considering a number of possible solutions to the loss of customer funds. These included a hard fork, or alterations to ethereum's code that would effectively reverse the hack, and a soft fork, which would enact code preventing the stolen funds from being redeemed. While live, approximately 10 million DAO tokens changed hands daily on the ethereum network.

## Other DAOs

Currently, a number of smaller DAOs have raised funds in ether or are in early stages of development, and trends in the market were beginning to take shape. Digix, a

DAO meant to create a gold-tracking asset for ethereum, raised $5.5 million in March in a crowdsale. MakerDAO, likewise, intends to launch a "stablecoin" with a fixed value that can enable a credit-based monetary system on the network. This formation, in which a team of developers raises money to deliver code that can then be managed by a diversity of participants, seems most common among entrepreneurs seeking to launch products or exchanges centered on ether trading or investing. Other notable projects that don't quite fit into this framework include Golem Project, which is building technology that would allow users to trade the idle time of their computers, and Augur, a decentralized prediction market.

# Chapter 8: A Step-By-Step Guide To Using Ethereum

At this point, you know essentially what Ethereum is. You know that you can't operate Ethereum without ether, and that an ether coin currently costs around $225. This next chapter is going to provide you with information regarding what you have to do in order to use Ethereum in a step-by-step fashion. After reading this chapter, you will know exactly the steps that you need to take in order to set up an Ethereum platform on your own computer.

Step 1 to Using Ethereum: Get Your Ether

Before anything else, the first thing that you need to do in order to start using Ethereum is to get yourself some ether. The easiest way to do this is to simply head to Gemini, Coinbase, or any other website that sells ether. Next, create a login for yourself and link your debit card to the account. After your debit card account has been verified, you will then be able to buy and sell ether at your leisure.

Step 2 to Using Ethereum: Download an Ethereum Mist Wallet

After you have your ether account setup, the next step is to download an Ethereum Mist wallet to your computer. It's important to make sure that you download the most recent version of the wallet code, so that you don't end up running into any unanticipated problems. Once your wallet is downloaded, you'll be able to send, receive, and store ether coins on your computer. Along with making sure that you're downloading the most recent version of code, you also want to make sure that you download the version of the wallet that is compatible with your computer. It wouldn't make any sense to download the Apple version of code when you own a PC. To find this code, a simple Google search for "most recent Ethereum Mist wallet download" or something along those lines.

Step 3 to Using Ethereum: Install the Ethereum Mist Wallet

Once you've found the code for which you're looking and have downloaded it to your computer, the next step is to allow the program to install. As it's installing, you are going to see a dialog box that resembles the following:

As you can see, there is a lot of information downloading to your computer. Patience is key while you're waiting for the node to finish its install. Once completed, you will then see the following screen:

While the previous screen was downloading Ethereum's nodes, you can see at the bottom of this screen that the blocks of Ethereum's blockchain will download at this point. At this point, you will also be prompted to choose whether you'd like to launch Ethereum with your own network, or whether you'd like to test the platform on a network where your personal ether will not have to be used. It's highly recommended that you choose to launch Ethereum through the test network initially. As you can see, if for some reason you have not purchased any ether for yourself prior to completing this step, you will still have time to do so after this step has been completed.

Step 4 to Using Ethereum: Create Your Password

As with most other types of accounts that you can create online, once Ethereum has finished downloading, you will be asked to create a password for your account. It's extremely important to recognize that once you choose a password, you will never be able to change it. For this reason, you should choose a password that is not only memorable, but also strong. You definitely do not want to put yourself into a situation where your wallet ends up being hacked because your password is too basic.

Step 5 to Using Ethereum: Become Familiar with Your Account Portal

After you've created your password, you will then be able to access your account's home page. This page is going to resemble the following image:

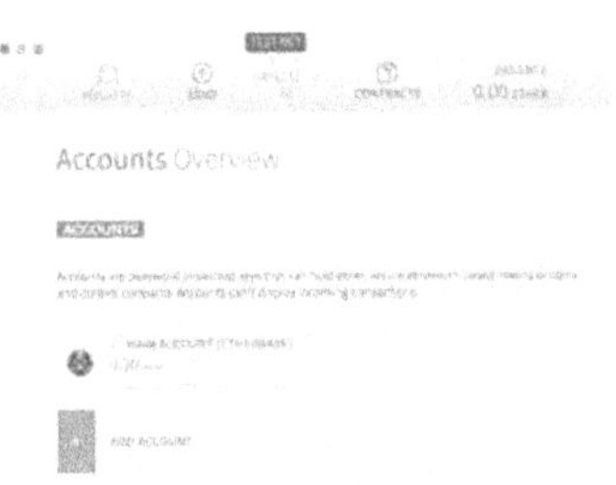


As you can see, your account will have your wallet, your contracts, and the option to send ether at the top of its menu. You also have the option of adding additional accounts to your ether profile under the same username that you've already created. This way, you're able to send and receive various types of currency, since you are able to turn your ether into any type of currency that you'd like. Remember, Ethereum allows you to choose the type of currency that you'd like to make your ether.

In the image above you can also see the number of nodes that Ethereum has accumulated in real time (this number is located below the red "Test Net"). If your Ethereum software stops working, you'll know because the number of nodes in the network will stop populating itself or this number will turn into a zero. If this does occur, you will probably have to restart the Ethereum application. This is a rather common problem that can occur, but it can be solved rather easily.

Step 6 to Using Ethereum: Understand Your Main Account Address

When you're working within any blockchain network, you're not going to have a username similar to when you're using an Instagram or Snapchat account. Unlike these types of applications where your identity is integral to the application's function, Ethereum attempts to protect the identities of its users.

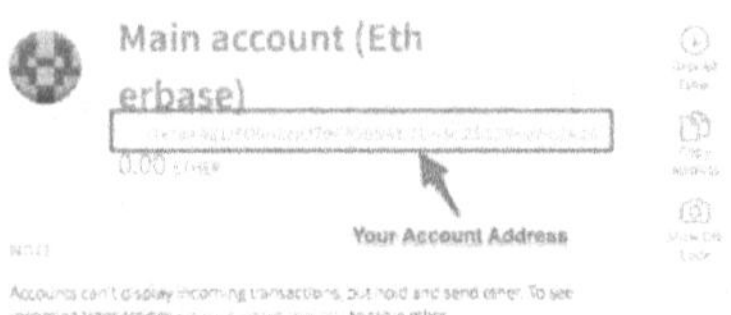


As you can see from the image above, your account address will be a long string of numbers, rather than anything that is easily identifiable.

Step 7 to Using Ethereum: Send Ether to Others

When you know the account address of someone else on the Ethereum network with whom you want to do business, sending money to them is incredibly simple. Click "Send" at the top of your account menu, and then type in the address of the receiver's account. Next, choose how much money you're going to give them. Keep in mind that you're going to have to pay a small transaction fee to the miners who are going to be processing this transaction for you. This will typically be only a small amount of ether, and your transaction should be completed within about thirty seconds. After deciding on the amount that you're going to be sending, you can hit the "Send" button. There is one last step. In order to completely finish the transaction, you must enter your account password. This helps to ensure that you are truly making the decisions for your account.

Step 8 to Using Ethereum: Look at All of Your Past Transactions

As with any blockchain-operated application, Ethereum has a public ledger that everyone on the network can see. If you want to see all of your previous transactions, all you have to do is to go to Wallet overview and then Latest Transactions.

The steps that were presented in this chapter should have been able to walk you through how to operate the Ethereum wallet and user interface. Hopefully, you now feel as if the process involved with setting up Ethereum and using it to trade currency with other people is rather straightforward. We have not yet discussed how to use Smart Contracts. That's because that will be the subject of the following chapter.

# Chapter 9: Most Important Cryptocurrencies Today

(Values based on the data recorded on 3rd January 2018)

Bitcoin

When many people hear the term 'cryptocurrency', they think of bitcoin but the truth is; bitcoin is just one of the many cryptocurrencies out there.

Created in 2009, many consider Bitcoin the first (working) peer-to-peer decentralized digital currency in the world. Therefore, it has the first-mover advantage.

It's currently the most known and recognized cryptocurrency, as well as the most widely accepted and widely used digital currency in many real-world transactions. Bitcoin has a market cap of about $251 billion and one Bitcoin currently has a value of $15,008, thus making it the most valuable and costly cryptocurrency on the market today.

## Altcoins

Altcoins come from the two words 'alt' and 'coin' which means that altcoins are alternative cryptocurrencies to Bitcoin. Therefore, any cryptocurrency other than Bitcoin is an altcoin.

Most altcoins are created on the basic framework offered by Bitcoin and most of them are essentially peer-to-peer; they also involve a mining process and provide inexpensive and efficient ways to perform network transactions.

Altcoins launched when Bitcoin showed success and potential after it its release 2009. Generally, they came up as alternatives to Bitcoin to solve and improve the features and protocol limitations of this giant (bitcoin).

All alternative currencies remain presently focused on becoming better substitutes to Bitcoin.

Here are a few more examples of such coins:

## Ethereum

Ethereum was proposed in 2013. Ethereum refers to a decentralized platform for apps running precisely as has been programmed without any chance of censorship, fraud or interference by third party. It offers a decentralized virtual machine called Ethereum Virtual Machine— EVM—meant to execute scripts through international public nodes network.

Even after suffering controversy that ended up in diverging blockchains, Ethereum is still one of the greatest platforms for smart contracts. Currently, its market cap is more than $85 billion and has a price of about $881. Critics say that from what is expected from digital currencies, the growth of Ethereum has been nearly negligible.

## Litecoin

Released in October 2011, Litecoin is a peer-to-peer, open source cryptocurrency that offers direct, almost-nil cost payments to any person across the globe. It is similar to Bitcoin in that users can mine it, use it as a currency, and also use it to pay for goods and services.

Litecoin has a market cap of about $13 billion and one Litecoin has a value of $250.

Dogecoin

Dogecoin is one of the most popular altcoins because it sort of started its life as a 'joke cryptocurrency' before exploding onto the market, harnessing the leverage of the online communities to fly high.

In a relatively short time, the currency has appreciated in value, with tons of users and internet communities all over the world sharing the cryptocurrency and memes about it online. Well, the hype did pay off because it's among the top cryptocurrencies on the market.

Dogecoin has a market cap of about $1 billion and one Dogecoin has a value of $0.009156.

Monero

Today, many often mislabel Bitcoin as the 'anonymous' currency, which is not the case. Monero, conversely, is a privacy-based cryptocurrency and in this regard, it uses the **ring signature technology**. It is not only private, but also secure and untraceable.

Down to eleventh place with a market cap of about $6 billion, folks who want to keep their anonymity on the web use it adamantly. Current value of monero is $429.67 USD

Ripple

Created in 2012, Ripple is a cryptocurrency and a payment network (RippleNet). Ripple links up banks and other large institutions and lets them transfer cash and certain assets through the network; all its transactions are logged on the XRP Ledger, which is decentralized.

By being a currency used in the payment network for the entire transactions, Ripple XRP reduces the time and money that usually goes into cross-border payments. Every transaction takes only four seconds to process through the system. Just to get a better perspective of this, consider that Bitcoin takes more than one hour, Ethereum more than two minutes, and traditional systems typically take up to five days to do the same.

Ripple has a market cap of about $144 billion and one Ripple has a value of$ $3.72.

Other popular cryptocurrencies today are:

IOTA

Dash

Zcash

## Chapter 10: Smart Contracts

The essence of what makes a smart contract **smart** is its ability to verify that the conditions set forth **within the contract** are met (or not met). Instead of a contract that is written by humans, and requires humans to interpret the results from the contract, smart contracts (once created and "signed") will automatically fulfill the contract's deal once the action(s) required are submitted. Since the contract is written in logical computer code, there aren't any subjective interpretations and fulfilling the contract triggers the contract's payment.

Smart contracts are capable of real-time data recording and communication through a nodal network. If you write a contract, for instance, that requires someone to watch a video, you could theoretically pay them by the second. With the proper coding in place, you would be able to see how many seconds they have watched as well as the constant flow of payment that they have received.

The thing is, you don't **have** to watch it, once written and deployed, the contract will fulfill everything outlined within the code as is possible.

The efficiency of these contracts is absolute. The cost to create them is minimal and reliant on the computational power required (in general, this will be small) the writing party will see a benefit of financial and time efficiency in comparison to the traditional routes. This is where larger players, such as insurance and finance companies will see a decrease in operating and overhead costs, as tasks required get outsourced or are given to employees that do not require a cubicle and infrastructure.

All tasks that require a person to verify if an action has taken place can now be distributed through the network through contracts and delocalized contract fulfillers, so long as the code is written correctly. To reiterate, here are the major reasons smart contracts will exist, and why a business would want to use them:

They can express business and financial logic as a computer program

They will represent the logical, triggering of events as messages to the program

Use digital signatures to prove the identity of who has sent the message

Everything is contained on blockchain (verified and logged).

Companies that use these smart contracts won't need to worry about liability issues for human errors that take place in nearly all business aspects. There is also less chance of intentional fraudulent activities (barring the issue with anonymity) because no one can lie about what has and hasn't been done. There is a portion of the Ethereum-contract network called Oracle. Oracle has the job of providing the data that is required for every contract to prove its performance while also sending commands and communications to the decentralized systems, contracts, and people where they are needed.

Oracle additionally verifies that everyone in the market and associated with any contracts is actually who they say they are. They must be accessing the contract from a secured point for Oracle to deem them trustworthy and for the data to proceed. There are inbound and outbound oracles that communicate with external applications and systems. They are required for smart contracts to do things such as release payments or to adhere to the details of every contract.

The inbound oracles give the smart contracts the external data that is relevant to the contract so that the contract can deduce what has been done and whether any aspects of the contract have been fulfilled. The outbound oracles communicate with internal command systems and are required for payment dispersal.

When you make and deploy a smart contract, an Oracle will need to be able to communicate in some way with the data enclosed within it which is otherwise not directly accessible to Ethereum. Only the most important data is going to be passed along from the external systems to the smart contracts that have been written. This is required because, in cryptocurrency, privacy and security are the normal, expected paradigm.

You can control the level of information that the Oracle has access to, such as **context** for the contract, leaving any information that is not paramount to execute the contract, completely inaccessible. This may be important when working on larger networks and with a large accessible population.

The benefits of smart contracts, I hope, at this point are obvious. It should be reasonable to believe that the benefits of these computationally coded contracts will increase dramatically in the future and that Ethereum is the best candidate so far for them to be executed with. This should begin to build a strong thesis for you to be investing (and probably using) Ether, Ethereum, and the smart contracts themselves. Since Ethereum is decentralized, the peer-to-peer nature of the blockchain gives superb verification and resilience. With that said, there are no third parties involved in a contract directly, other than its verification through Oracle and the DOA.

Being able to read the code of a contract will be something required if you are going to be writing or fulfilling contracts. Once you have a handle on the coding language, however, everything will be logical, and you can write efficient programs (the more efficient, the more likely to be fulfilled and the cheaper to make/execute) as well as read them. What should be noted is that once a contract is fulfilled, it's most likely completely solidified as a finished deal. Imagine writing a contract that requires the person accepting to hand you 100 Ether. If they sign that contract, they will pay you 100 Ether, and you will receive it. Depending on the privacy of the contract, that person will have no access to any of your information at all, and no way to dispute any problem. The inverse, of course, is also true, so being able to read logical programming language will be essential for taking on contracts as well as writing them!

As the owner of any contract you write (or buy), you will be required to fill the payment details if the contract is to be fulfilled. How to write your own contract will begin with downloading Solidity, which is a programming language similar to Javascript. Another option would use the programming language that is web based, called Cosmo. Cosmo has a benefit of being able to be sent directly to the network. Accessing the contract requires another program called Ethereum Web3.ja. With either Cosmo or Solidity, you'll still need to follow the same steps outlined below. Without following these steps to completion, you may risk a contract that malfunctions or is not able to be uploaded to the network at all.

Once in the network, your contract can be distributed to all of the necessary application frameworks. This will make it much easier for you if you prefer not to have to go through everything when it comes to actually uploading your contract. There will also be less human error possible when done this way. The framework is there only to guide you; you will still need to have some input and work done for everything to work properly. Thankfully, there are many ways to debug possible errors in contracts and make sure they work properly before uploading them into the network as final projects. You should only deploy perfectly working contracts, or you will risk possibly serious issues, such as paying for an unexecuted, or only partially fulfilled contract!

There are many frameworks to use; some of the most popular are Embark, Truffle, Meteor, and API. You should read some on each to make sure they suit your needs.

Once you have all the programming and frameworks to build your smart contract, the actual steps to creating it are highlighted below. Just remember that each contract will be different based on who is writing it, as well as every person fulfilling a contract may also execute the solution differently.

You will need to create an Ethereum node to write the contract on. To do this, you'll download that program touched on earlier called Geth. Geth is an accessibility program for the main interface which Ethereum nodes are implemented within, and Ethereum itself uses. Other programs that are needed will be on the Ethereum website (www.ethereum.org), or they will correctly link you to trusted, required sites. Make sure that the sites are trusted!

Once downloading of your application is complete and it launches correctly, you will need to compile a smart contract with the use of Solidity or Cosmo as referenced earlier. These programs will execute your program, or contract, and tell you if it is working properly as you expect it to. No worries if something isn't to your liking, you can always edit your contract at a later time as long as it has not been deployed.

After finishing the contract, you can deploy it. To deploy a contract, you will have to spend Ether to have it “hosted” on Ethereum so that others can access it and sign it. You will also sign the contract to let the network know you are the owner.

Now you will obtain a blockchain address and API for your contract. You **can** call your contract back by use of the API, but you will most certainly want to try to work everything out before deploying the contract. Also realize that you could be spending Ether every time you put up or take down your contract, or interact with your contract!

Creating the contract and putting it onto the Ethereum network requires that you test it to make sure it will work as it did in the debugging stage before deploying. Ethereum itself has a basal expectation of all contracts hosted on the network, so you will need to make sure that you fall into those correct assumptions. While being the sole owner of the contract, you are the one who determines the end goals of the contract. However, Ethereum itself will need to look and make sure your logical code is correct and communicable across the network. There are a few steps to make this process easy and irrelevant for future contracts you pursue.

Testing your transaction times is essential because you'll need the network time for verification. If you set the transaction time too low, a savvy "hacker" may be able to swindle more money from your contract than you initially intended by sending hundreds of commands per second. This doesn't give enough time for everything to be verified, and as such will cause the network to lag in response to you submitting payments, which could amount to more than you anticipated for! You should be able to code, into your program, a way to defend against this type of "attack" but to be safe, make sure there is always a delay for sending payments that are at least 10 seconds. This will give enough time for the peer-to-peer network to verify what is going on. You may want to set up a basic “trial” contract to make sure you understand the process as you move forward.

Use Solidity to be able to access your Ethereum nodes (contracts). You also need to keep your Python library separated from the virtual environment that you are going to be working in. This ensures that you’re not using local environments for network environments, and you’re also not placing your contract in a place you didn’t intend.

You’ll need to start with a new client node with your console window. Begin with Truffle or the other framework applications mentioned in the new window and deploy your contract with the “truffle deploy” command. This is a boilerplate contract, and the program is going to automatically detect obvious bugs with this while additionally testing transaction times. Just remember that 10 seconds is a good amount.

Once deployed, it will be beneficial to run a compile in your framework program such that you make sure your contract is accurately compiled to your specification.

If your program is not at risk of potentially being lost, stolen, or draining your account (or any other danger), you can deploy your program onto the Ethereum network to debug there. This is more expensive and riskier, but will most accurately predict if there are issues you need to resolve.

Once the contract is completely debugged and you feel safe, it's time to actually deploy it (finally) to the network.

In Truffle, you'll specify "truffle init" for a new directory which you can specify.

Identify your contract to put within the directory.

Open config/app.json and add your contract to the contracts area that is provided by Truffle.

Restart your node in a new window and run the command "tesrpc".

Now you will be ready to run the root directory, verify the contract is on the network in a manner you wish it to be.

Now you can test run using Truffle in the root directory and make sure all tests pass. Once sufficient to your liking, you can add the UI to the Truffle Directory and run Truffle such that the UI is automatically compiled and created into the contract and it can be created in any directory.

Recompiling the contract regularly is a good idea to assure that the application is running correctly and all changes made are reflected within the Truffle program. The App directory is where the boiler points are housed and will help with the UI as well as the distribution for the contract you've written. If access is needed, you have to start the Truffle Watch process and then reopen your root directory in the browser window. Now you'll need to open the developer and do a right click to choose the selection you need to inspect. On this screen, you should add the "window.onload" function so that you can ensure the contract is actually activated when your page loads.

Make sure all your functions are copied over and that any testing assertions are removed, and that output is returned to normal using your console. Here, all you should need to do is load Meteor and create your UI so that it is easily used for interaction between anyone and your contract. The better the UI, the simpler and easier the interaction will be.

# Chapter 11: The Current And Futures Uses Of Ethereum

Because Ethereum has become so popular, many people have started to look towards potential avenues for which it can be used in the future. In the meantime, as well, it's adopted many different uses that take advantage of the multifaceted nature of this amazing technology.

Perhaps the most interesting of these are the DAOs and DApps. These were originally outlined in the Ethereum whitepaper and since have taken on many different forms. In one specific instance, there was a DAO related meltdown that caused the course of Ethereum to change forever.

## DAO

So what exactly is a DAO? DAO stands for **decentralized autonomous organization**. So what is it?

Well, a decentralized autonomous organization at its core is the idea that something, somewhat like a company, can be hard-coded and leaderless. The notion is that the decentralized company can be self-sustaining through the people using it without some sort of leadership at the head of it. This can dramatically affect the way that the company is run.

In essence, a decentralized autonomous organization is exactly what it sounds like. It is decentralized, which means that it is distributed amongst multiple people instead of being concentrated in the hands of a central figure or entity. It is autonomous, which means that it contains the logic to run on its own and be heavily, if not entirely, automated. And it's an organization, which is self-explanatory.

The most popular DAO of all time was **The DAO**. The DAO was a catastrophic project that would end massively changing Ethereum. So what was The DAO?

The DAO built upon the idea of decentralized autonomous organizations by implementing it. The manner in which it was implemented was that it was a bundle of different smart contracts. These smart contracts were developed from the very beginning, and everybody who linked up to the service would take advantage of the infrastructure that was preemptively developed.

What would happen is that people would invest in the service and then obtain DAO tokens. These tokens could be used by the people holding them to vote on different projects. The funds used by the investiture into DAO tokens would then be allocated towards investment in that given project.

It is with this that the principle behind decentralized autonomous organizations becomes extremely clear. The notion is that they will allow the users to all take part in one mutually funded entity. In essence, The DAO was a gigantic crowdfunding effort. However, it also served as a proof-of-concept for an amazing technological idea by implementing an infrastructure for a decentralized autonomous organization.

Unfortunately, The DAO would meet a poor fate. While the infrastructure was incredible, there was a fatal security flaw that allowed it to be intruded. In late 2016, The DAO was hacked.

To understand why this was a big deal, you need to under the sheer depth and size of The DAO. The DAO was, and remains, the biggest bundle of smart contracts to date. The project was of a sheerly unthinkable size, and it was worth millions of dollars of investiture at the time of the hacking.

However, over a long period of time, many Ethereum users had their investments into The DAO taken away from them, with the hacker stealing more than a million dollars worth of Ether. The DAO quickly crumbled as a result of this intrusion, but that still didn't solve people's lost investitures.

So what happened? Well, the Ethereum community was fairly well split on the issue. The obvious solution would be to simply roll back the blockchain. However, there was one key issue to this:

blockchains aren't really **designed** to be used in such a way. They're actually designed to be pretty much immutable without a massive effort.

While the massive effort was certainly possible, the fact that it was even being considered was off-putting to many Ethereum users. The idea of immutability and not tampering with markets is a huge driving force behind the success of cryptocurrencies, which had largely been picked up by libertarians who saw the decentralized currency as a means of emulating a perfectly untamperable market which would thereby be undistorted.

Therefore, when people were seriously considering the notion of tampering what should be untamperable, there was a huge amount of backlash. Some people weren't even against it for political reasons; some people simply didn't like the idea of having to redo transactions that had taken place or otherwise losing currency, effectively, during the period between the The DAO hack and the rollback.

The Ethereum founders would act somewhat immediately on the The DAO hack by trying to recover and restore funds that were invested in The DAO. However, there were still a huge number of lost funds that were outright taken during the attack. These were going to be impossible to recover. This first solution would be referred to as a **soft fork**; this is when an arbitrary change to the blockchain is made without the intent of rolling it back forthright.

However, more needed to be done to correct the incident that had taken place. The Ethereum community would put the issue to a vote. The issue at hand was whether it was better to roll back the blockchain to prior to The DAO's hack and then restore funds from there, or whether it was better to leave it as it was.

The people in favor of the former were, by and large, those who felt that the intrusion should never have occurred in the first place and was little more than a fluke. They, understandably, wanted their lost money back and were going to do whatever they could to ensure that such would happen.

The people in favor of the latter were those who saw the blockchain as immutable. It was also composed largely of people who saw the incident as a natural result of people investing in an unstable technology. They didn't think it was fair to tamper with things that are natural consequences.

In the end, the former won out. The Ethereum blockchain was rolled back, and the vulnerability was fixed. The DAO wouldn't last much longer after this incident, however.

This also, coincidentally, led to one of the biggest events in Ethereum history: the development of an alternative Ethereum. The people who were put off by the idea of changing the blockchain were very much not happy to see it be changed. Since Ethereum is open-source, they simply started another fork of Ethereum. This Ethereum fork would be known as Ethereum Classic.

Ethereum Classic isn't the focus of this book, of course, but it is a fundamental and foundational part of the Ethereum canon because it would take many of Ethereum's users. While Ethereum has continued to grow long since the accident, there are still quite a lot of people who have opted instead to use Ethereum Classic, which is for all intents and purposes just Ethereum but with the original blockchain. Where Ethereum is currently second in terms of market capitalization, sitting only behind Bitcoin, Ethereum Classic is sitting somewhere around eighth. This is quite a feat when you consider how much of a dent in the cryptocurrency market really is.

In the end, it's really easy for the story of The DAO to turn you off of decentralized autonomous organizations, but it shouldn't. The truth is that it was just a negative implementation of an infinitely useful innovation. The potential of decentralized autonomous organizations is insane.

So what went wrong with The DAO? Since the idea of a DAO is that it's, well, decentralized, it's almost impossible to change the code once it has already been deployed to the Ethereum blockchain. This means that the code will run pretty much infinitely. However, if there were to be a bug in the code, there would be almost no way for one to fix the bug.

So what can be done to ameliorate the situation? Nobody is quite sure. The best bet is just to write extremely secure code and not deploy anything that isn't well-tested and within a high degree of certainty to not crumble in the same way that The DAO did. This, however, is far easier said than done. This doesn't, however, mean that such would be unfeasible.

**DApp**

DApp just stands for **decentralized application**. These are things which run on the Ethereum blockchain that carry out certain operations in a certain manner by running consistently.

In essence, there are just smart contracts that are organized into decentralized applications that anybody can use. The vagueness of this definition provides very well for something with a lot of potentials.

Such strange and peculiar things have been enacted through the use of smart contracts! It goes far beyond anything that I'm sure you're thinking of at this point. Let me outline a few of the most interesting ones. All of these really push the boundaries of what Ethereum and smart contracts can do.

Take, for example, Etheria. Etheria is based off of the hit game Minecraft. What it is is essentially a voxel world which is divided into little tiles. You can buy these partitions for one Ether. However, once you're in ownership of a partition, nobody else can touch it, and you can send commands to your little section of Etheria using smart commands. This allows you to build and add whatever you want. While it's still relatively in its infancy, by going to the Etheria website, you can zoom in on the Etheria world and see little horses and structures that people have built.

It's also impossible to make it through the DApp section without bringing up EtherTweet. EtherTweet is heavily based off of Twitter. It allows users to send messages which are stored on and retrieved from the blockchain, all of which are limited to 160 characters. This appeals to people who are against censorship because the only people who can remove a message on this application are the people that post them. Remember, this is a large part of the appeal of decentralization: the fact that no one person is in control and, therefore, no one person may tamper with things.

Perhaps one of the most disruptive applications that may be popping up is TenX. TenX, if it carries out, will allow people to use their Ether coins at anywhere that they can use Visa or MasterCard cards. They can, therefore, be used anywhere that accepts Visa or MasterCard by being automatically converted to the local fiat currency. This would be disruptive for obvious reasons: it promises that a cryptocurrency can hit the mainstream and be used day-to-day instead of remaining a niche mainly segregated to the internet.

## Future Uses

As you can see, Ethereum shows an infinitely expanding number of use cases. It also shows the best potential to be picked up as a mainstream cryptocurrency.

What does this mean for you as a potential investor? Well, it means that even though it may already be a bit of a pricey coin, it's only going to grow bigger. Your best time to get into Ethereum is **right now**. Just get into it and then sit on it. Why is this?

Well, Ethereum's development team is one of the most forward-thinking development teams in the cryptocurrency realm, and they also have enough of a market share that they could easily start to overtake the market.

So if you're ready to take the plunge and buy some Ethereum, head on to the next chapter.

# Chapter 12: Ethereum Classic Vs Ethereum

In the summer of 2017, when the DAO (Decentralized Autonomous Organization) was hacked, the hackers exploited a loophole in the DAO. However, the ether could not be taken out of the system because the smart contract was set in a way that any invested ether that was subject to a withdrawal was not accessible for 28 days. Therefore, the ether community had time to act, with three plausible solutions: **Do nothing**, **a soft fork** or **a hard fork.**

Now, we shall not get into the complexities of the two forks. However, when it was agreed upon that subsequent action would be to conduct a hard fork, there was a huge uproar and split amongst the ether community with a small sect being a staunch anti-hard fork. Those who were opposed to the hard fork refused to move to the new blockchain and thus stayed back and remained in the old blockchain and named it, **Ethereum Classic** or **ETC**.

And with that, we are now in the thick of the ether war between ETH and ETC. The war is not only ideological but also ethical and has been termed as the most important moment in the entire history of cryptocurrency, second only to the birth of Bitcoin.

So, what is Ethereum classic (ETC) then?

As just previously mentioned, ETC is the product child of the resistance of the hard fork and was by virtue of semantics, the original ether. As of the time of the writing of this, one coin stands at 11.35 USD with the market cap estimated at 1.08 billion dollars and it is currently the 5th most expensive cryptocurrency in the world.

But wait, why did some people feel the need to stick to the old blockchain while all the big players, including the founders Gavin Wood and Vitalik Buterin, made the move to the new chain? One simple answer, **philosophy**. When cryptocurrencies were introduced, Ether included, the cryptos were meant to serve one core purpose which would have been to eliminate financial corruption; a system that was not susceptible to human whims. Therefore, to the ETC community, a hard folk would be entirely contrary to what cryptocurrencies stand for, and it would offer a convenient cop-out. By hard forking the system, this entirely meant that it was being altered by human ability.

Then again, what is Ethereum Hard Fork (ETH)?

As already stated, this was a new blockchain that resulted from the hard fork, hence the name Ethereum Hard Fork, and it is considered the new Ethereum. As of the writing of this article, one ETH coin stands at 279.09 USD with the market cap a mind-bending 26 billion dollars and is second most expensive cryptocurrency in the world, after, of course, Bitcoin.

Sadly, however, to the ETC fanatics and sympathizers, ETH is the new Ethereum with all the big hitters moving to it. And unlike what most detractors might think, the new blockchain was formed for one reason, and one reason only; to ensure that the money stolen from the DAO hack was, in fact, returned to its rightful owners. ETH stands for so much more, a victory for the ether community. On the brink of the worst cryptocurrency hack, the community stuck together and came up with a product stronger and better than its predecessor.

Now that we are all on the same page let us take a deeper look at some of their key differences including the pros and cons of either.

Major Setbacks

**ETC**

Ethereum classic faces one major setback, the lack of backward compatibility. Everyone moved to ETH meaning that ETC cannot access any updates whatsoever that are conducted by ETH. For instance, ETH moved to the new PoS (Proof of

Stake) and PoW (Proof of Work) and sadly the ETC community is short chained when it comes to these and other updates.

However, Classic's biggest problem is a conspiracy. A lot of people tend to believe that ETC could be an attempted attack on Ethereum itself. They reasoned out that during the whole attack and split period, anti-Ethereum folk joined the ETC camp just to bring about chaos and disruption amongst the ether community. There have been a lot of references and analogies concerning ETC with a famous financial blogger terming the chain as an insecure orphan child.

**ETH**

As mentioned before, the main drawback of ETH is that how it was set up goes against its only core value which is to end financial corruption and avoid and possible human interference. Some people felt that the hard fork did exactly that.

Another setback, but in line to hard forks is, the detractors weren't entirely convinced that in future other hard forks were not a possibility. Since ETH can be subjected to hard forks, this would mean that it is entirely within the realms of possibility that other hard forks will occur, causing many different versions of ether and subsequently devaluing the cryptocurrency.

However, this is unlikely to occur since a hard fork requires a majority vote from the community for it actually to occur. Nonetheless, it is also still possible for the community to vote for it and it shouldn't be brushed off as unthinkable.

Let us now look at the Pros and Cons of each.

**ETC**

**Pros**

- The currency is immutable and stays true to its philosophy.

- A number of big industry players have of late being backing it hence increased the potential for it to do better.

**Cons**

- Does not receive updates because it is not backward compatible.
- All the heavy hitters already moved to ETH
- Many people view it as an insult and attack on Ethereum
- Scammers have made their way into the ETC community.

**ETH**

**Pros**

- It is growing disruptively and exponentially.
- Constant updates ensure that it works better.
- It has to back from a majority of the original bigwigs and many more are joining in.
- The stolen money was reversed and returned to its rightful owners.

- ETH has a better and higher hash rate than ETC
- Represents what happens when people stick together for a common course. This will be vital and prove key in the event that similar problems and/or new ones arise in the Ethereum community.
- ETH has a powerful corporate backing of around two hundred companies called the Enterprise Ethereum Alliance (EEA) which include players like Microsoft, Toyota, JP Morgan, ING and many more.

**Cons**

- It went against its policy core value of immutability.

So, where should your loyalty lie?

To each their own, right? Wrong. What would then be the purpose of this discussion if it lacked a bit of subjectivity and expert advice as to which side of the fence you should be sitting on? Having analyzed the key differences and pros and cons of the two versions of the currency, it is rather obvious which side you should pick in this war. However, being objective, I would urge you to pick a side you feel is in line with what you stand for. Nonetheless, before you hop onto the ETC train by integrity, here are two compelling arguments that you should keep in mind.

**Immutability:** While we cannot sit and deny that ETH went against policy, the circumstances and situation need to be taken into account. When faced with such a situation, drastic measures have to be deployed. Often, even in our own lives, we are caught out in situations where the solution goes against our values. The hard fork, hard as it was a pill to swallow, did a lot of good for the Ethereum community. The hacker's amount was greatly devalued, and everyone was refunded. So, win-win.

**Possibilities of other hard forks in the future:** often, people have bashed democracy as a system that is flawed. Well, not in the ether community. This is because it is decentralized and very democratic. This means that major decisions never boil down to one individual or a small group. If such critical decisions have to be made, a majority vote to be considered with the whole community voting. Therefore, the possibility of other hard forks or human whims for that matter is greatly diminished.

Ethereum has faced and beaten the biggest setback any cryptocurrency has faced since their inception. It turned and an absolute disaster on its head and is now flourishing with as many people predicting that it will soon overtake Bitcoin as the largest cryptocurrency in the world. It is also predicted that Ether will also only be the second cryptocurrency to break the 1000 USD mark and with the backing of the EEA, it can only grow.

The future is certainly bright for ETH.

# Chapter 13: What Is Ethereum Mining?

With all the technological advances in computing, the whole world seems to be going digital. Every industry must adapt to these changes or be left behind in the technology race. But one goal remains the same - to maximize profit and sustain development. It is this goal to make life easier which gave birth to the cryptocurrencies. Ethereum is one of those emerging cryptocurrencies and it has advantages over the others. Some even consider it to be an improved version of its older and bigger brother, Bitcoin. Like Bitcoin, Ethereum can also be mined using pretty much the same method and hardware setup. But how is it done and is it profitable to do?

Mining cryptocurrency like ether is an intensive computational work that requires time and processing power- lots of both. In mining, you participate in consensus with a cryptocurrency network as a peer. You are then rewarded for solving complex mathematical problems. Basically, you run a mining application which puts your computer to use.

As previously explained, all information processed during cryptocurrency transactions are embedded in the data blocks. Each of these data blocks is internally linked to other blocks which form the blockchain. The data blocks need to be analyzed as quickly as possible in order to ensure smooth transaction flow inside the platform. This requires processing capabilities that are beyond what the currency issuers can provide. This is where mining comes in.

A miner is also an investor in some sense. But instead of buying and selling cryptocurrencies, he invests computer power, electricity, and time while sorting through the data blocks and analyzing them. When the right harsh is hit during the mining process, the solutions are submitted to the currency issuers. The issuer then verifies the solution, and then offers parts of those transactions they have verified. They can also offer cryptocurrency as payment to the miners' work. The outcome of mining is referred to as proof-of-work system. Some cryptocurrencies use only this system while others may use proof-of-stake in addition to proof-of-work.

It's called mining because it's like mining for precious metals such as gold or platinum. Like precious metals, digital currencies are hard to come by. But don't think that it's a way to get rich quickly. It will eat up a lot of your time and effort. This is particularly true when you're working alone. There's a major difference in that traditional mining increases the amount of precious metals available. Digital mining increases the amount of cryptocurrencies in circulation.

This principle also applies to Ethereum. Ethereum can only be utilized through the mining outcome. But Ethereum mining goes beyond increasing the amount of circulating ether. It also secures the whole Ethereum network as blocks inside the blockchain are created, verified, and subsequently published.

Ether is the fuel that makes the Ethereum platform run smoothly, making it essential. Look at ether as additional payment given to software developers for creating topnotch applications.

The supply of ether is finite. The total amount ether to be ever released was decided during the presale in 2014. Every year, only 18 million ether will be released. This reduces inflation possibilities.

Before consensus validation, a given difficulty's proof-of-work must be provided for each block. This validation algorithm is called Esthash. The difficulty is dynamically adjusted to ensure that on average, 12 seconds is required to produce one block in the network. This can't be altered except when the miner owns more than half of the computing power in the Ethereum network.

How To Do Ethereum Mining

You can do Ethereum mining in the comfort of your home. You'll need to learn some script writing and a bit of knowledge of the command line. It's an exciting and easy practice.

1. Set up your mining equipment ready. There are quite a few Ethereum mining calculators on the Internet. You may want to check how much profit you will earn using your current computer rig.

Note: Ethereum mining takes a lot of electricity. To offset the increase in your utility bills, the mining should be efficiently carried out so you get income by selling the ether you mined.

Any computer, desktop or laptop, can be used for Ethereum mining but it has to have a graphics processing unit or GPU card with 2GB of RAM or more. The computer's central processing unit or CPU alone will not lead to profit because of the high operations cost.

GPUs, on average, are at least 200 times faster. This comes in handy during

Ethereum mining. AMD GPU cards also seem to be more efficient than NVidia ones.

Here are the basic system requirements for a mining computer setup:

- Operating system (OS) should be 64-bit versions of Windows or a Linux flavor.

- The mainboard should have enough PCI-E slots depending on the number of GPU cards. Some setup will require getting a PCI-E riser for each GPU added.

- The power supply should have enough connections to power the GPUs installed on the system. Total system power draw should also be computed beforehand to determine the wattage of the power supply needed. There are sites that can do the computing for you.

- A low-end CPU should be enough as long as it has 4GB of RAM or more. Make sure the mainboard, CPU, and RAM are compatible with each other. Check the specifications of each component before buying.

- For input and output, you'll need a mouse, a keyboard and a monitor.

Unlike Bitcoin, mining Ethereum using a simple home computer setup is still profitable because the latter doesn't employ ASICs for mining. In its current state mining Bitcoin with the same rig will result in negative profit.

2. Once you get your mining hardware set up, you'll need an Ethereum wallet to store your Ether. You may opt to get Ethereum's official wallet but it will take time since you need to download and sync the whole Ethereum blockchain. Or you may use Ethereum wallet sites like MyEtherWallet to skip the download process.

Those who are not familiar with the DOS command line may download the user-friendly Mist. This package already has an Ethereum wallet where you'll receive mining profits. Mist has an Ethereum browser built-in where you can chat with other miners and get valuable tips.

3. Once you have installed Mist, you'll need to wait until the Ethereum blockchain has finished downloading and syncing. It's currently above 10 GB in size so it will take a while depending on your Internet bandwidth. Spend some time familiarizing with Mist.

When the Ethereum download has finished, you then need to setup your wallet. Open it and generate an account and a wallet based on the contract. Your wallet will hold your payout address where you can receive the rewards from the mining activity.

4. The only component missing now is the mining software. There are quite a few Ethereum mining applications and the Claymore Ethereum Mines is considered as one of the more popular clients out there.

There are useful tweaks that you can do to your Windows computer before you start mining. By default, Windows will have your computer go into 'Sleep' mode when it's idle to conserve power. This stops the mining process so you need to turn this feature off in Windows Control Panel. Setting the system page file to 16 GB will add virtual memory to your setup for handling larger data. Disabling Windows Updates can also prevent your computer from restarting by itself and ending the ongoing mining. You also need to configure your antivirus software to exclude "EthDcrMiner64.exe", Claymore's executable file.

You can do solo mining of Ethereum but it will only be good as a hobby. To obtain significant profit mining Ethereum on your own, you'll need more than a couple of GPUs. Try a building full of them. And as the saying goes, 'If you can't beat 'em, join 'em'. Joining a mining pool is a more practical and profitable scenario. You join a mining pool and share the profit with the members of the pool. Due to the sudden rise of its popularity, a lot of mining pools have sprouted overnight. Etheremine and Nanopool are two of the best ones.

5. Once you have everything set up, it's time to start mining. When you run your mining software, it will initialize the GPUs in your computer, create the DAG file for the GPUs and begin hashing. With mining pools, like Ethermine, you can always check your earning in real time.

## The Future of Ethereum Mining

In its current implementation, Ethereum, like Bitcoin, uses the proof-of-work system. It refers to the solving of advanced mathematical equations which the miner is required to clear before he can add his block to the blockchain. There is environmental damage incurred due to electricity demands but there is profit to be achieved through Ethereum mining. There are mining calculators available to help you determine how much you will earn and if it is even practical to dwell in Ethereum mining given the current scenario.

In the future, there are plans to eliminate the mining process and replace it with an algorithm designed for consensus. This will be included in Ethereum's next update called Serenity and it will adopt the proof-of-stake system over proof-of-work.

The Ethereum is a complex connection that is maintained by a network of computers. The impact of these computers is undeniable and there is still profit to be gained for the ether miners. However, this will soon change and mining will come to a stop when the consensus algorithm replaces the current system. But at the moment, everyone is advised to continue mining and think about the problem later.

# Chapter 14: Effect Of Blockchain And Ethereum On Economy

There are a few components of blockchain computing, such as clients, services, applications, platforms, storage and infrastructure.

Services Provided by Blockchain

SaaS (Software as a particular Service)

It is an important model to back web services along with SOA (service-based architecture). It is good for new developing approaches like Ajax. There are different SaaS service offering services via the internet, such as email, CRM, office suite, etc. it is hosted on a scalable infrastructure and can be accessed by an ordinary web browser.

PaaS (Platform as a Particular Service)

It is an integrated and abstracted service to support management, running and development of other applications. It is helpful for developers to scale their apps without worrying for infrastructure.

IaaS (Infrastructure as a Particular Service)

It is a way to deliver servers, storage, space, memory, infrastructure and bandwidth with the self-service console without the help of an IT team.

HaaS (Hardware as Service)

The user can lease the hardware for his personal use and this option enables you to save money on the maintenance of equipment. It can be deployed on your own infrastructure and appropriate software.

Practical Applications of Blockchain Computing

There are a few practical applications of the blockchain computing in the business organizations and different fields:

Blockchain Computing in Business

It offers better hardware, software and network resources to provide unique services on the web and servers. It can identify and meet the logical demands of business organizations and offer innovative services. Organizations currently use traditional infrastructure and enable users to increase the benefits of IT resources in data centers. Companies get the advantage of the traditional management of data centers and practice the use of IT resources for an end user. It can be done in various steps, such as procuring, finding floor space and offer sufficient cooling and power systems. A blockchain can make it easy to implement the automation, workflow of business and abstraction of resources. You can add a shopping card on the website for the convenience of users. This process can increase the efficiency of resources and more than one person can use it at one time.

## Blockchain Computing and Education

It is important for educational institutions to quickly respond to the needs of students. They can cope with the fixed budgets and staff. Any challenge in the field of education can be easily handled with the help of blockchain computing.

With the help of blockchain computing, education services are reliable and economical in the fast-growing industries of this world. It is good for research, collaboration and discussion. It enables you to run classes at different remote locations and institutes.

Online Entertainment

Lots of people come on the internet for entertainment and these people can get the advantage of blockchain computing. The entertainment based on the blockchain can easily access TV, mobile, set-top box and other forms. You can get the advantage of the quality of time and better clarity. You can search for (ODE) on demand entertainment, such as news, games, videos, and audio. Amazon, Netflix, YouTube, and Hulu are internet giants and they are earning good profits into the entertainment industry. You can solve ODE puzzles and offer a solution to different issues regarding entertainment.

## Blockchain Computing and Telecommunication

Numerous telecommunication companies use blockchaining computing for public and private networks for domestic and commercial purposes. Blockchain communications are voice communications, database communications, and applications hosted by the third-party other than an organization. These can be accessed via the public internet and these communication services help you to manage better relations. You can use a smart phone, tablets, and other third party devices to increase the productivity of your business. These services will be over the support of VoIP system deployments and collaboration for conferencing system. These can be accessed from different locations and you can manage your business with them.

## Banking and Finance

With the growth of the international market, there is no need to have a separate database for client and a separate portal for banks. For a faster and advanced business, it is important to offer reliable access to customers to a few major concerns of banks. Financial institutions are quickly using blockchain-based services to increase the agility and reduce the cost of ownership. They offer consolidation, virtualization, data center and storage for easy disaster recovery. The blockchain computing is becoming mature and reliable to increase the adaptation of this system.

Service Provider of Blockchain Computing

There are a few major service providers of blockchain computing to facilitate all customers in a better way:

Google 101-Network

It is based on millions of economical servers that are used to store data, such as numerous copies of World Wide Web. You can search millions of queries at a faster rate.

Microsoft Azure

It is an internet-scale service used as MS data centers and offers a unique functionality to build applications from the web to enterprise scenarios.

It is a web service interface that offers resizable capacity to blockchain for computing. It is designed for developers to easily use their actual capacity.

IBM BlockchainBrust

It is designed for regular users and it offers blockchain computing service with the IBM blue service software.

## Chapter 15: Ethereum Keyplayers And Technical Infrastructure

The Ethereum infrastructure has broad distribution. We have discussed the computers called nodes, among which are the miners, and their essential role in the Ethereum blockchain. There is also an Ethereum community dedicated to the success of this massive project.

A very important person still deeply involved in Ethereum, is Vitalik Buterin, whose white paper in 2014 started the Ethereum project. During the short life of Ethereum some key players have left, one such person was Gavin Wood lead developer in the C++ computer language. His replacement was Christian Reitweissner, creator of the Solidity computer language, expressly created for smart contracts. Other key personnel is Taylor Gerring who is the director of technology; Martin Becze who is the head of JavaScript client developments; Fabian Vogel Steller who is the head of dApp development; Alex Vander Sandy, who is the head missed developer and Viktor Tron, who is the leading Swarm developer. Swarm is heavily involved with artificial intelligence.

Despite these leading lights, the development of Ethereum code is a community effort. An Ethereum Foundation is leading the development of Ethereum. It is a not-for profit company headquartered in the city of Zug, Switzerland, mentioned before. The foundation has a governing board whose executive director is Ms. Ming Chan and whose members are Buterin himself, Jeffrey Wilcke and David Ben Key who, unlike Buterin and Wilcke, is not a professional programmer but rather a lawyer.

The rollout of Ethereum is not yet complete it is currently on Metropolis version 3, with Serenity version 4 planned for next year.

# Chapter 16: How To Make Money Trading Cryptocurrency

With just a few dollars worth of Bitcoin you can start trading cryptocurrencies right now. There are no broker fees, there are no middlemen to deal with, nor really any barriers to entry or red tape. All you need is some percentage of a single Bitcoin. There is no reason not to try it out. If you can accept risking a few dollars, it's a great way to get into cryptocurrency.

I started trading with less than $40 worth of Bitcoin.

I gradually traded my way up to 5.5 Bitcoin (worth over $5000 at the time) in less than a month or two. This isn't to suggest that trading is something that's easy or effortless. Losing money is an inevitable part of trading and investing, but you can certainly minimize risk and loses with the right strategies. The reality is that if trading were an easy, risk free way to make money, everyone would be a trader. However, if you're a strategically minded person, patient, and able to research and analyze market trends, you'll enjoy trading cryptocurrency.

Cryptocurrency is the real Occupy Wall Street

Being a decentralized ledger, the Blockchain can never be controlled or manipulated by a single institution. Its design makes transactions virtually error proof, and it can also do much more than just transfer the ownership of digital currency; it can be used for transferring assets and shares of companies, smart contracts, commodities, and escrow services. This technology will likely change the future of finance as we know it, democratizing financial markets while simultaneously eliminating "banksters."

If you're just getting your feet wet with cryptocurrency, all the technical jargon can seem overwhelming.

It's important to learn, but for now, If you're just interested in trading and investing, having a basic common-sense understanding of business, consumer demand, and economics is enough to give you an edge over other traders (at the moment). Most of the current batch of traders are early cryptocurrency adopters, cryptocurrency "miners," programmers, and basically people that are more tech savvy than business/market savvy.

They're focused on small technological innovations that help build hype for a coin in the short term, without giving much thought about how the coin will exist outside of the exchanges and crypto community. This gives you a huge advantage.

So let's get started. First buy some Bitcoin.

There are some exchanges that will let you purchase specific cryptocurrencies for USD, but it's a better idea to buy Bitcoin first. With some Bitcoin, you can trade into and out of every other cryptocurrency on the market, on every crypto exchange. Remember: you don't have to buy a whole Bitcoin ($390 as of writing this); you can purchase Bitcoin in fractions known as

Satoshis; for example, 500k Satoshis equals 0.005 Bitcoin. The safest, most popular place to purchase Bitcoin is coinbase.com, however you can also go to an exchange that has a USD-BTC pairing to try to trade USD for Bitcoin at a cheaper rate.

Now that you have some Bitcoin, it's time to find an exchange.

The most reliable exchange I've found is Bittrex.com. There are other exchanges: some are good, some are bad, some have been shut down already — the Mt. Gox scandal might ring a bell. Some people are discouraged from cryptocurrency altogether when there's news of an exchange getting shut down or coins being stolen, but I see all of this as a right of passage for any new market that is still in its infancy. I find it very encouraging that most of these shady exchanges have been terminated and their CEOs have been doxxed and sued to hell.

News spreads veryquickly in the crypto world, so check news feeds daily.

You will usually see smoke before there's a fire, as long as you pay attention to the news on twitter. Crypto exchanges and businesses are being talked about on twitter. Check in on twitter and crypto forums daily, follow hash tags, see what people are talking about. Information is power, news is power, and rumors are opportunities!

Once you have Bitcoin in your exchange account, you can start trading.

However, before you just randomly pick some cryptocurrencies and watch their charts, I recommend you do some research first; otherwise you're trading blindly. The best way to learn about each coin is to search it, like "Cannabiscoin ann" – "ann" as in announcement. This search phrase will lead you to the bitcointalk.org forums, to the official announcement thread of Cannabiscoin.

Trading basics.

Researching the market is referred to as "fundamental analysis." By gaining the right information at the right time and understanding how it will interact with the market, it becomes easier to stay predict trends — essentially whether or not a cryptocoin will rise or fall. In addition to fundamental analysis, you also have "technical analysis." Technical analysis is equally important, but it refers specially to studying charts and finding patters—for example, at a certain price, a coin will fall repeatedly.

The best time to buy a coin is after it has been dumped.

Why? Because the people that didn't cash out during the pump (called "bag holders") don't want to sell their coin at the bottom, at a much lower price. It goes without saying that if the price of a coin you've bought moves upward quickly, it's best to cash out, back into Bitcoin. And If it's a good coin that you want to invest in for the long term, make sure you buy back in after a dump. Sometimes it is better to focus on accumulating good coins rather than making more Bitcoin, because a good coin will always rise again.

# Chapter 17: Ethereum And The Future

As the second most traded digital coin in the world, Ethereum's future is very promising. Just how far can it go? No one really knows the answer to that question. One thing that we can all be assured of is that to date, it's upward mobility is expected to reach an all-time high in the very near future. In fact, it appears to have unlimited potential that will be tapped in the years ahead.

#21. Be prepared for new things in the future: As a platform that has opened the door to an unlimited number of uses, Ethereum, may be the catalyst that will literally change the way the world does business. While no one knows exactly what to expect we can tell you what Ethereum enthusiasts are saying that we should be looking ahead to.

Flippening

Flippening is the day that many Ethereum investors are anxiously anticipating. The day when Ethereum actually overtakes Bitcoin as the highest traded crypto in the market. On that day, Ether will become the number one coin in the world.

Is it possible? No one knows, but we do know that no other cryptocoin has even come as close as Ether. It is literally the only coin in position to even challenge Bitcoin. Of course, there are a lot of things that must happen before that day, but as it stands right now, Ether is ready to do it.

Bitcoin itself is viewed by the world as a currency. As such, it is in direct competition with many of the government issued currencies. As a result, many of these governments may put restrictions in place that could literally curtail Bitcoin's growth giving Ethereum a chance to catch up.

Ethereum however, does not compete with any powerful entities and kind of blends into the background. By not putting itself out in front and vulnerable to attack, it may be able to avoid the legal issues and regulations that could one day be imposed on Bitcoin. The end result could be that Ethereum may be a safer investment option for the future in comparison.

Ethereum's Fuel Dynamic

Because Ether is traded more like a commodity than a currency, it is expected to continue its appreciation in value. Sometimes referred to as digital oil, its innovative platform is expected to revolutionize the world in much the same way as the printing press made revolutionary changes hundreds of years ago.

There is definitely a rush expected that could be compared to the gold rush of the 1800s. As more people learn about Ethereum and what it can do, many are anticipating a digital oil rush for the 2000s.

Proof of Stake Algorithm

The Proof of Work protocol was the first introduced on the Bitcoin blockchain. It has proven to be successful time and time again. Ethereum's venture to stray from that proven formula has also stirred up heavy anticipation for its growth in the future. By using the Proof of Stake protocol, the role of the miner will evolve into being more like a prospector staking his claim on a certain territory. They will "hold" to their interests and will use their Ether not just as a buy, sell, or trade option but as a means to verify or "stamp" transactions on the network.

This could be a highly lucrative opportunity where investors or those who hold a stake in Ether will actually be paid to maintain the network rather than the miners. This means that in time, the role of miners of Ether will eventually fade away and investors themselves will be the ones verifying transactions and keeping the system running.

This alone could stir up excitement in the world of Ether. Without the use of miners, there will be no use for major shell-out of cash for expensive equipment. No longer becoming a part of a mining pool to get a return on their investment. A simple Ether purchase and a stake will be all that is needed. An important factor to consider if you're interested in mining for Ether.

The Enterprise Ethereum Alliance

The association of companies that are working together on a battery of experiments that will help them tap into all the many abilities of Ethereum is another reason the coin has such a positive future. Their mission is to connect Fortune 500 enterprises, startup entities, academics, and technology together via the Ethereum platform. The group hopes to learn from and even enhance Ethereum's Smart Contracts and use them to define even more enterprise-grade software to create new applications that can speed up the business process. With companies like BP, Cisco, Credit Suisse, HP, Intel, Microsoft, Stanford Law School, and the US Coast Guard supporting it, all the makings of a new era are on the horizon. With these companies taking the lead, others are sure to follow suit making Ethereum the only cryptocurrency right now that has such an impressive collection of partners pushing it forward.

While we don't know exactly what can happen in the future, there is an awful lot of speculation about Ethereum's potential. Whether it will reach the heights of Bitcoin remains to be seen, but at the same time, all the pieces are in place for Ethereum to have a most promising future.

# Chapter 18: Investing In Ethereum

Part of the massive popularity of cryptocurrencies in 2017 is Ethereum. It is a decentralized system, which disrupts the need for third-parties or banks in sending payments across the internet.

The total market value of cryptocurrencies increased to more than $200 billion. These alternative currencies even entered the mainstream with the introduction of the first Bitcoin Futures on the Chicago Board Options Exchange. So, here are some reasons why you should invest in Ethereum.

1. It is growing in popularity

In a span of two years, Ethereum managed to establish itself as one of the fastest growing digital currencies in the market, second only to the largest digital currency in the world. The idea of Ethereum was first proposed by a 19 year old, Vitalik Buterin, in 2013. Presently, millions of people have already obtained this currency, though many say that the idea of having third-party apps that can run on their network is its main attraction.

2. Ethereum's value is rising

Ether, the valuation of Ethereum's currency, has gradually increased these past months as people have started setting up their own cryptocurrency wallets. It has been used for public trades since 2016 and the release of Ether is only limited to eighteen million every twelve months. These days, it isn't surprising at all to see a single Ether being valued at more than 700 USD.

3. Ether could be the future of currency

Ethereum and other digital currencies have the capacity to innovate and transform the financial system, in the same way Airbnb and Uber have done in their own field. Confidence in conventional markets is low given the emergence of a financial crisis ten years ago, and people are now becoming more confident with what the online world has to offer.

4. You can trade on exchanges

It's a good thing to have options as there are numerous cryptocurrency exchanges that allow Ethereum on their platform. One of the most distinguished, Coinbase, faces competition from exchanges like the Buy Virtual Currency. This new platform offers individual account managers and effective customer service – all while accepting almost every digital currency available.

5. You can already use ETH to pay for products and services

More and more businesses are starting to accept digital currency as payment. Also, as people become more comfortable with cryptocurrencies, it's to be expected that the number of organizations that will get involved with Ether will rise over the next few years. In fact, Overstock.com, an online retailer of furniture, bedding, and DIY, which is located near Salt Lake City, notified their customers that they are already accepting cryptocurrencies as payment.

# Chapter 19: Mining Ethereum

While Bitcoin and Ethereum both exist because of Blockchain technology, there are a lot of differences that need to be understood. If most of your education about mining revolved around Bitcoin, then you're going to find something entirely different here.

One of the first things you'll discover is that Ethereum runs on Ether, so when you choose to invest, you're actually buying Ether, not Ethereum. For the miner, Ether is the reason he works on the Ethereum network.

While Ether is a form of currency, it is used in a very different way than Bitcoin. With Bitcoin, you can spend or trade it in a variety of ways, but Ether can only be used on the Ethereum platform. When clients use a smart contract on Ethereum, they pay for this service with Ether. From the miner's point of view, it is the means of greasing the wheels of the network. It is also the form of payment a miner will receive. So, when you are mining on the

Ethereum platform, your ultimate goal is to acquire Ether by validating transactions on the network.

Why Do It?

There are a number of reasons why you might want to mine Ether rather than just buy it outright. For each person, that choice is different. It is based on your personality, investment style, specific circumstances, and your financial position.

No one can tell you that you should or should not mine Ether. It is a decision you must make for yourself. However, there are several common reasons why people might want to consider taking on this type of endeavor.

First and foremost is the money. For every successful block, you add to the Blockchain you are paid in Ether. If you're looking for a way to get into the market, this might be it.

How much you make will depend on your success rate at solving the blocks. You will have to decide beforehand if you feel that you can make more from mining Ether than you can from making an outright investment. It is important that you factor in the cost of the equipment you will have to obtain as well as the cost of power. Please note, while it may cost a lot to get set up with your mining rig, it is a one-time cost. After you've started, it is just a matter of how successful you will be in mining those coins.

Another reason you might want to consider mining Ether is to give support to the network. The more miners in the system, the more reliable it will be. This will naturally cause the price to go up, and you'll make a profit.

People also mine as a way to accumulate more coins over the long term. As the price of the coin continues to rise, so will the value of the Ether you've earned. It could end up being a pretty significant bottom line for you and a huge boost to your personal net worth.

I'm sure that you can come up with plenty of other reasons for mining Ether that haven't been discussed here. The main idea though, is that it will serve as a means to boost your personal net worth, to increase your income, and give you personal satisfaction that you are supporting a whole new economic system.

## How to Get Set Up

Unlike other cryptocurrencies that require a very specific collection of hardware to mine, Ethereum's mining can be done on a wide variety of platforms. Even someone with a home computer can have a successful attempt at mining Ether. That's if their home setup meets certain specifications.

The biggest obstacle that Ethereum miners face is how to balance the cost of powering their equipment with the value of the Ether they receive. When you're new to the mining process, the best way to make money is by joining a mining pool. However, you still have the option of setting up your own mining rig yourself and working independently. In this chapter, we'll discuss both options so you can have a pretty good idea of how well each one works.

Hardware Needed: Since Ethereum does not use the same consensus algorithm as Bitcoin, the hardware must also be different. If you've been mining for Bitcoin in the past, you should know that the ASIC hardware you have will not work with the Ethereum network.

This is good news because it opens the door for more people to mine that do not have the money to lay out for that kind of expensive equipment. To set up your own mining rig, here are the things you will need:

A motherboard: This is what will make it possible for all the different components to communicate with each other.

A Graphics Card: This is the part of the system that processes the consensus algorithm.

Adequate Storage capacity: You will need to have enough storage space to download the Blockchain and all the transactions that have been previously verified.

Memory: You will need to have at least 8 GB of memory and possibly more. The files on your system will be consistently growing so whenever possible, opt for the higher memory.

Power Supply Unit: This will generate the power you need to run your rig.

Ethernet: Your mining rig will need to be connected to the Ethernet to process transactions correctly.

Of all the components needed to set up your rig, the Graphics Card is the most crucial. If you want to increase your chances of making money, you can invest in more than one Graphics Card. The more you have, the better your chances of being able to solve a block. It is important to recognize though, that the more equipment you have, the more power you will need to expend. So make sure that it balances out. It might be best to start with one card and add later after you have found some level of success in the endeavor.

For the most part, mining Ether is primarily done on the equipment you have. When you have a computer that can run 10,000 different possible permutations, it is easy to see that it is difficult for a solo miner to compete even under the best of conditions. However, there are a few things that can make your equipment run more efficiently, which could increase your chances of finding the right solution to the blocks that need to be mined.

Download Geth. This is the application that will allow you communicate with the Ethereum platform.

Unzip the file and transfer it all to your computer's HDD. In most cases, this will be the C: drive.

Execute the installation

On the command terminal, type in ‘cd/’ into the command prompt.

Create a new account on Geth.

Create a password, write it down and store it someplace safe.

Let Geth link up with the network. This is the action that will begin the download of Ethereum’s Blockchain and sync your computer with the global network.

Download the mining software. Ethminer is a favored one on the Ethereum network, but there are others.

Install the software

Repeat step 4

In the new terminal window, type ‘cd prog’ and tab.

Go to the Ethereum mining software folder and type in ‘cd cpp’ then tab and enter.

To begin mining, type in ‘ethminer-G’ and press enter. This step will start the mining process.

These are simple basic setup procedures. However, there could be some variations depending on the type of equipment you have. Once you are all setup, test out the system to make sure that you are getting the most out of your equipment.

How to Do It

Getting the equipment setup and synced with the Ethereum network may take some time and it will require you to have at least a basic knowledge of how this type of hardware works. However, if you have some background in this area, it shouldn't be a difficult process. Just make sure that you follow the guidelines for the type of equipment you purchased so that you reduce the odds of complications from using incompatible components.

Once you're set up, and everything is communicating well with the network, you are ready to get started.

The idea of using the term "mining" came into play because of the comparison to the

19th-century gold mining. While a miner's primary goal is to validate transactions on the Ethereum network, the work they do also produces new Ether that can be used in the system or later traded for other cryptocurrencies or exchanged for traditional fiat currency.

This happens when the miner is able to figure out the unique hash of a block. A block is a collection of data that details the information pertaining to all the transactions completed within a specific period of time. Every new block created will have a unique alphanumeric string that has to meet certain characteristics.

It must connect with the previous block

All characters within it must be interrelated.

It must contain a combination of alphanumeric characters along with symbols.

The detailed math behind this process creates a hash that works like a digital fingerprint for each block, no other block will have the same hash.

The idea of making money from mining can be exciting, but mining is not as simple as it sounds. It is not as if you're given a set of blocks and told to decipher these hashes. Instead, it works more like a competition. The details of the transactions are dropped into something called a pool, and the miner must take them from there. The challenge is that there could be hundreds of miners who are working on the same block. The one who is rewarded is the one that solves it first. This means that it is often the case that you are working on a particular block and another miner solves it. At that point, you must go back to the pool and begin the process all over again with another block. As a result, there could be days or even weeks before you are able to solve a single hash and collect your Ether.

Miner's serve very specific roles on the network. Aside from confirming transactions, they are the key players that protect the network from those who would like to cheat the system. They make sure that every transaction that is approved is authentic. They are also on the front lines protecting the network from numerous forms of cyber attacks, and they keep the massive decentralized machine functioning.

Types of Mining

There are several different types of mining setups you might want to think about. The first is what is called a home farm, which consists of all the mining equipment located in a single location. This type of mining is very expensive, as it requires the miner to setup and pay for an entire mining rig themselves. The good news is that the equipment necessary for mining Ether can be used for other purposes so if later on you decide that you want to give up on the practice, it has some resale value.

Another type of mining is referred to as cloud mining. This type is for people who do not want to spend a lot of money to set up an entire mining rig in their own home. Instead, they can mine remotely and simply share the processing power.

Caution is warranted when working with cloud mining; because you are not in control of the entire process, you are also not in control of the expenses you will incur when choosing to mine. The equipment you use is not your personal property so you may find you have to pay out more than you bargained for.

If you choose to use cloud mining, there are a few things you need to be aware of.

Be wary of any group that promises you huge profits

Make sure that you will have access to technical support when you need it

Check them out thoroughly before you join. You can read reviews and monitor any chatter you may hear in the mining forums

Always make sure that you follow whatever security protocols they have set in place.

Never let anyone know where the mining farms are located.

Mining equipment can be costly. For this reason, not very many people choose to make money mining. It is not only expensive to acquire the needed hardware, but it is also expensive to run it. This makes it a vulnerable target for thieves and others so if you choose this route, it is necessary that you maintain a certain level of anonymity to protect your investment.

Is It Worth It?

Mining is not for everyone. Those who do well in this endeavor are those who can comfortably haggle with suppliers, and has enough computer knowledge to put the equipment together themselves. To make the most out of your equipment, you want it to last a long time. Here are a few suggestions that could extend the life of your equipment.

Adjust your core voltage usage by setting it below zero. Try dropping it all the way down to -100. That will reduce the amount of energy your system will use when mining.

You can also lower your core clock to bring down the temperature. This will also save on the wear and tear of your equipment.

By applying overclocking at system startup, you won't have to go back and set it again after your system starts running.

Because a miner relies heavily on the efficiency of their equipment you need to do all that you can to extend its life. Probably the greatest enemy to mining equipment is heat, so anything you can do to lower the temperature and reduce power usage will not only help your equipment to work better, but it can also make it last longer.

The ability to make a success of mining is dependent on two things: the cost of the hardware and power. Returns are not consistent, and they are dependent on the price of Ether at any given time. Because of the extreme fluctuations in the market, a certain amount of Ether one month could be worth a great deal but the next month could be significantly lower, or higher. Still, if you have the right setup and a good energy source, it is possible to make a pretty good income from it.

There are several ways you can enter the mining system and the decision as to whether the cost of getting set up is worth it depends entirely on you. Weigh the costs of your initial investment in equipment, the amount of time you will have to dedicate to it, and the other costs and then make your decision.

## Chapter 20: The Future Of Ethereum

The Blockchain technology exposed the world to the concept of trustless Information structures, providing us a glimpse of the near

future with the launch of the Bitcoin crypto currency. The Bitcoin crypto currency system has really altered the currency

landscape. However, its key challenges have forced key developers to rethink its model and focus on the more open source, scalable

and advanced Ethereum technology.

Certainly, the Ethereum technology has provided us a Glimpse of a better future. Iteratively, it's a remarkable step forward,

however we aren't there yet. It was possible due to this Ethereum's underlying technology. However, the current consensus

algorithm--the PoW--has triggered the growth of Ethereum technology.

The shift in PoW to PoS in the future could significantly Reduce the computational load of the network, making Ethereum technology

more efficient and efficient. The distributed storage alternatives as well as also the state channels could also substantially

increase its capabilities. Right now, the solutions supplied by the omnipresent Blockchain technology as well as also the Ethereum

system are still under development. Therefore, they may not come for their pictured fruition.

· As competition amongst Organizations in many businesses stiffens, the Blockchain technology is likely to Become even more

in-demand later on.

While there are some competing Blockchain programs on the market, together with the Backing of the tech-enthusiasts and companies, it stands to reason that the Ethereum job can become the go-to as more companies seek to include the Technology in their own

operations.

# Chapter 21: Ethereum Mining

In this chapter, you are going to learn what Ethereum mining is, how it works and what you need to become a miner. This chapter will also look at the profitability of Ethereum mining.

## What Is Ethereum Mining

In the course of this book, I have mentioned the terms mining and miners, and you might be wondering what the terms mean. To understand what mining is, let us take a look at how transactions happen on the Ethereum blockchain. When you send some Ether to another user, your wallet broadcasts the details of the transaction to the Ethereum network. This transaction is grouped together with other pending transactions to form a block. Nodes within the network then try to confirm the validity of the block. To do this, these nodes must solve complex mathematical equations. The first computer to find a solution announces to the network that it has found a solution. While the other nodes in the network do not know the actual solution, they can confirm whether it is the correct solution by passing it through a hashing function. If more than 51% of the computers within the network confirm that the solution is correct, the block gets added to the blockchain. This system of achieving consensus is known as Proof of Work (PoW), since the node adds the block by proving to the others that it has worked to solve the complex mathematical equations. The node that found the solution is rewarded with newly released Ether, as well as the transaction fees for the block. This process of validating blocks and generating new Ether is what is known as mining.

The mining process is very computationally demanding. The nodes competing to find solutions to the complex mathematical solutions are essentially computers with high processing capabilities that are programmed to continuously run a hashing algorithm. The process of validating blocks within the

Ethereum network takes about 12 seconds. After a block is added to the blockchain, the process starts all over again. By expending your processing power to mine on the Ethereum network, you can profit from the block reward (about 5 Ether) and the transaction fees awarded to you every time your computer mines a new block.

One of the key differences between Ethereum mining and Bitcoin mining is the hashing algorithms used by the two networks. Whereas Bitcoin uses the SHA256 algorithm, Ethereum uses an algorithm known as Ethash. Ethash makes it more effective to mine Ethereum using ordinary GPU's, unlike the SHA256 algorithm, which is more suited to ASIC's (Application Specific Integrated Circuits). ASIC's are very expensive, something that has made Bitcoin mining the preserve of the elite club of those who can afford ASIC's. Ethereum mining, on the other hand, is more decentralized, since GPU's are a lot more affordable.

Ethereum Mining Hardware

During the early days of the network, it was possible to mine Ethereum using an ordinary laptop or desktop PC. However, as more and more people started mining, the difficulty rate of mining Ethereum went on increasing to the point where it became impossible to mine using an ordinary PC. Today, if you want to get into Ethereum mining, you have to invest in the hardware to build a mining rig that has enough processing power.

To build a mining rig, the first thing you need to invest in is about six or more GPU's, the kind that are used for 3D video games. These GPU's should have a 3GB capacity or more. While it is still possible to use CPU's for mining, GPU's are more efficient since they are optimized for repeatedly running similar operations. Next, you need to find a motherboard for your mining rig. The motherboard should have at least 6 PCI slots for attaching the risers. You also need a power supply unit that will comfortably handle all the GPU's running simultaneously. The kind of power supply will be determined by the number of GPU's you decide to use on your rig.

The next piece of hardware you need to get is some powered risers. If possible, go for those that come with all capacitors built in. Risers will allow you to connect all your GPUs to the motherboard simultaneously. You also need to find a CPU and some RAM. However, since your mining rig won't be handling any multitasking tasks, the CPU and RAM need not be expensive. Finally, you will need a hard drive (about 60 GB or more will do), fans for cooling your rig and an Ethernet cable. You should not use your mining rig with Wi-fi.

## Ethereum Mining Software

After you have assembled your mining rig, you will also need to install the software that will allow it to interact with the Ethereum blockchain. The first thing you need to do after building your rig is to install the drivers for your GPU. These can easily be downloaded from the manufacturer website. Some manufacturers will even provide the drivers alongside the GPU. From there, you need to install the mining software. There are several mining softwares that you can use for Ethereum mining, with the most popular being ClayMore Dual Miner.

You will also need to configure your rig as a node. This process will connect your rig to the Ethereum network and download the blockchain on your rig's hard drive. To configure use a tool known as Geth. If you want a simpler option, you can go with Ethermine or MinerGate. Once you finish configuring and connecting your rig to the Ethereum blockchain, you are ready to start mining. You can also perform other operations on the network, such as writing and executing smart contracts and building decentralized apps.

Pool, Solo Or Cloud Mining?

Now that your mining rig is ready, and your miming software has been set up, you are ready to become a miner. However, before you can get started, you need to decide how you will be mining. Ethereum mining usually takes one of the following three forms:

Pool Mining

This is the easiest and the most effective way to get started with Ethereum mining. Pool mining involves teaming up with other miners. Each of you provides their processing power to the mining operation. With the combined processing power, there is a higher chance for the pool to find block solutions than when you are working alone. This is a great way to ensure that you keep earning steady returns from your mining operations. Pool mining is also a lot easier for beginners, since there us a pool admin to help you solve any of the initial challenges you might encounter. However, you also need to keep in mind that mining as a pool means that you will have to share your rewards with the other members of the pool. The likelihood and frequency of finding block solutions will depend on the block size.

While I recommend pool mining, this does not mean that you should join the first pool you come across. Different pools have their advantages and disadvantages. Before joining a pool, there are a number of factors you need to consider. The first factor is the pool size. This is because the more the members within a pool, the higher the hashing power within the pool, which in turn means that the pool has a higher chance of finding block solutions. However, this also means that there will be more people to split the rewards with. All the same, it is a lot better to join a bigger pool. While your rewards per block will be certainly lower, you will be assured of getting regular rewards.

The second thing you need to consider is the minimum payout offered by the pool. The minimum payout refers to the smallest amount of Ether you need to accumulate before your payment is sent to your wallet. The larger the minimum payout, the longer you will have to wait before receiving your payments. I don't find this to be a good arrangement. Ideally, you should opt for pools that have a very little minimum payout. A small minimum payout means you will receive your payments more frequently. This gives you the flexibility of being able to quickly leave a mining pool if you feel that it is not the right one for you.

The next thing you need to think about is the pool fee. Being a member of a pool does not come for free. You have to pay a regular membership fee. These fees go to the pool administrator, since running a pool is basically a full-time job. The fees are also used for the expenses of running the mining pool. Pool fees are usually calculated as a percentage. The standard pool fee lies between 1% and 4%. Sometimes, you will come across some mining pools that have 0% fees. Such pools are usually supported by donations. However, they are not the most reliable, so you should opt for those with a fee of about 1% or 2%.

Finally, you need to consider the payment structure offered by the mining pool. The payment structures used by most mining pools can be divided into two major categories:

**Proportional system**: With this method, all the members receive a share of the block reward based on the percentage of hashing power they contribute to the pool. The more the hashing power you provide, the higher the rewards you receive.

**Pay-per-share system**: With this system, the payments do not depend on the number of blocks mined by the pool. Instead, the pool administrator calculates the number of blocks the pool is expected to mine in a certain period, based on the laws of probability. From this number, the administrator comes up with a fixed number of Ether that is paid to the members regularly. This means that, with the pay-per-share system, members receive steady returns whether the pool has mined blocks or not.

Solo Mining

If you do not want to mine as part of a mining pool, you can always go it solo. Mining solo means that you do not have to share your returns with anyone. However, you need to keep in mind that mining is a competition. All the miners are trying to be the first one to find the block solution so that they can win the block reward. You are in competition with all the miners, including those who have combined their hashing power together by forming mining pools. Therefore, as a solo miner, you have a very small chance of mining any blocks.

For you to succeed as a solo miner, you need access to hundreds of GPU's. Not only is gaining access to such resources wildly expensive, but you will also need to handle other problems. For instance, you have to build dedicated cooling systems for your multiple mining rigs to ensure that they don't break down from overheating. The noise from the rigs and dedicated cooling system also means that you cannot run such an operation from your apartment. This means having to get a warehouse or garage for your mining operation. In addition, your electricity costs would be extremely high. To avoid all these challenges while still running a profitable mining operation, it makes a lot more sense to join a mining pool.

## Cloud Mining

If you want to mine Ethereum without having to handle the actual mining operation, you can opt for cloud mining. This is where you pay someone with the right equipment to do the actual mining for you. In other words, you are hiring their mining equipment and mining time. Initially, this might not look like a very logical arrangement. After all, why would someone who has invested in mining equipment mine for you instead of mining for themselves? However, cloud mining has a number of advantages to both parties. Since you pay for cloud mining services upfront, this is a way for the service provider to regain a guaranteed profit from their investment in mining equipment. On the other hand, paying for cloud mining services allows you to pass the cost of maintaining mining equipment to the service provider. For instance, if a mining rig breaks down, it is not up to you to purchase another one. You also do not have to build cooling systems in your home for your mining equipment or live with noisy equipment in your apartment. While cloud mining seems convenient, mining for yourself is more profitable, therefore I do not recommend cloud mining.

## Can You Make Profits From Ethereum Mining?

So, is there money to be made in Ethereum mining? If you have the right equipment, you can make money from mining Ethereum. It's also good to keep in mind that the profitability of mining will also depend on some variable costs, such as the cost of electricity and equipment maintenance fees. Mining is an energy intensive operation. The lower you can keep your electricity costs, the more profit you will get from mining. The profitability of Ethereum mining also depends on the mining difficulty. The more people get into mining, the more difficult mining will become, which will lower miners' profitability. All in all, Ethereum mining is not a get rich quick scheme. However, you can make a modest amount of money per month from Ethereum mining.

It is also good to keep in mind that there are plans to shift the Ethereum network to a Proof of Stake (PoS) consensus system. This new system will not require nodes to solve the computationally demanding mathematical equations. Instead, they will be required to stake a percentage of the Ether they own in order to validate transactions. As the shift to the PoS consensus system approaches, Ethereum mining will become more difficult, which will in turn translate to less profits from mining. Once the shift is implemented, that will be the end of Ethereum mining, since the PoS system does not depend on mining to keep the Ethereum network running.

# Conclusion

Thank you again for downloading this book!

The investment process for Ethereum should seem very comfortable by now. Everything you have read in this book is meant to prepare you and give you a beginner's basis in Ethereum. The book is intended to make you understand the foundations of how Ethereum came to be in the simplest way possible with less technical jargon. They say a good foundation builds a safe house, and the same goes for investing. If you do not fully understand what your product is, and how it shifts on the market, then you will fail at trading or using it to your benefit.

If you wish to invest further in Ethereum, then one of the main things to do is to follow the trends. Just as it is in regular trading, you must keep up with the news on the crypto-currency that you have put your money in; in this case, Ethereum. You can buy all the Ethereum using your life savings and end up broke in less than an hour. This is because it takes more than just two days of trading to get it right. It takes time and patience. Don't rush into your investment.

www.ingramcontent.com/pod-product-compliance
Lightning Source LLC
LaVergne TN
LVHW011937070526
838202LV00054B/4693

*9781990373657*